Microneurosurgical Anatomy and Surgical Technique

Xiang'En Shi • Long Wang • Hai Qian
Editors

Microneurosurgical Anatomy and Surgical Technique

中国科学技术出版社
CHINA SCIENCE AND TECHNOLOGY PRESS

Springer

Editors
Xiang'En Shi
Department of Neurological Surgery
Beijing Sanbo Brain Hospital
Beijing, China

Long Wang
Department of Neurological Surgery
Beijing Sanbo Brain Hospital
Beijing, China

Hai Qian
Department of Neurological Surgery
Beijing Sanbo Brain Hospital
Beijing, China

ISBN 978-981-19-8275-0   ISBN 978-981-19-8273-6 (eBook)
https://doi.org/10.1007/978-981-19-8273-6

Jointly published with China Science and Technology Press
The print edition is not for sale in China (Mainland). Customers from China (Mainland) please order the print book from: China Science and Technology Press.

This Springer imprint is published by the registered company Springer Nature Singapore Pte Ltd.
The registered company address is: 152 Beach Road, #21-01/04 Gateway East, Singapore 189721, Singapore

Contents

Anatomy of Cranium, Dura Mater, Brain Surface, and Surgical Techniques

Xiang'En Shi, Long Wang, and Hai Qian

1 Anatomy of Cranium

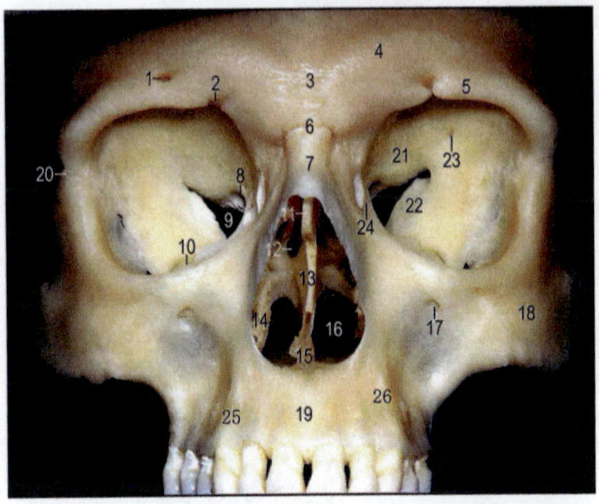

Fig. 1 Anterior view of cranium. (*1*) Supraorbital foramen, (*2*) Supraorbital notch, (*3*) Glabella, (*4*) Superciliary arch, (*5*) Supraorbital margin, (*6*) Frontonasal suture, (*7*) Nasion, (*8*) Optic foramen, (*9*) Superior orbital fissure, (*10*) Inferior orbital fissure, (*11*) Perpendicular plate of ethmoid bone, (*12*) Aperture of sphenoid sinus, (*13*) Vomer, (*14*) Inferior nasal concha, (*15*) Anterior nasal spine, (*16*) Nasal cavity, (*17*) Infraorbital foramen, (*18*) Zygomatic bone, (*19*) Maxilla, (*20*) Frontozygomatic suture, (*21*) Lesser wing of sphenoid bone, (*22*) Greater wing of sphenoid bone, (*23*) Lacrimal foramen (Hyrtl foramen), (*24*) Lacrimal groove, (*25*) Alveolar process, (*26*) Maxillary process

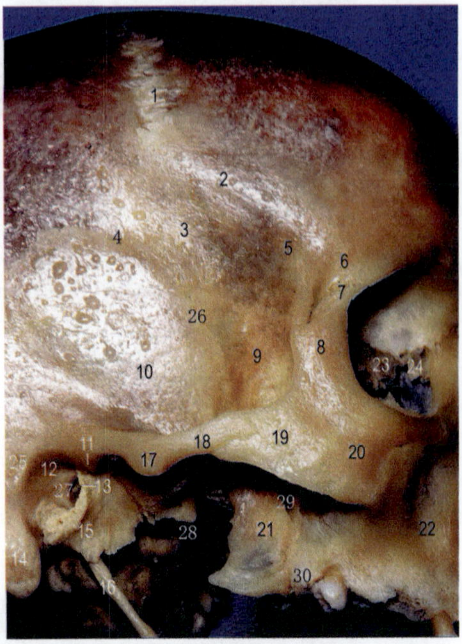

Fig. 2 Lateral view of right frontal, parietal, temporal, and zygomatic regions) zygomatic parts). (*1*) Coronal suture, (*2*) Superior temporal line, (*3*) Inferior temporal line, (*4*) Squamosal suture, (*5*) Temporal surface of frontal bone, (*6*) Zygomatic process of frontal bone, (*7*) Frontozygomatic suture, (*8*) Frontal process of zygomatic bone, (*9*) Greater wing of sphenoid bone, (*10*) Squamous part of temporal bone, (*11*) Postglenoid process, (*12*) Suprameatal triangle, (*13*) Anterior margin of external acoustic meatus, (*14*) Mastoid process, (*15*) Tympanic part of temporal bone, (*16*) Styloid process, (*17*) Articular tubercle, (*18*) Zygomatic arch, (*19*) Temporal process of zygomatic bone, (*20*) Zygomatic bone, (*21*) Lateral pterygoid plate, (*22*) Maxilla, (*23*) Palatine bone, (*24*) Lacrimal bone, (*25*) Suprameatal spine, (*26*) Pterion, (*27*) External acoustic meatus, (*28*) Clivus, (*29*) Pterygomaxillary fissure, (*30*) Maxillary tuberosity

X. Shi (✉) · L. Wang · H. Qian
Neurological Surgery, SanBo Brain Hospital, Capital Medical University, Beijing, China
e-mail: shixen@ccmu.edu.cn; wanglong@ccmu.edu.cn; qianh@ccmu.edu.cn

X. Shi et al. (eds.), *Microneurosurgical Anatomy and Surgical Technique*,
https://doi.org/10.1007/978-981-19-8273-6_1

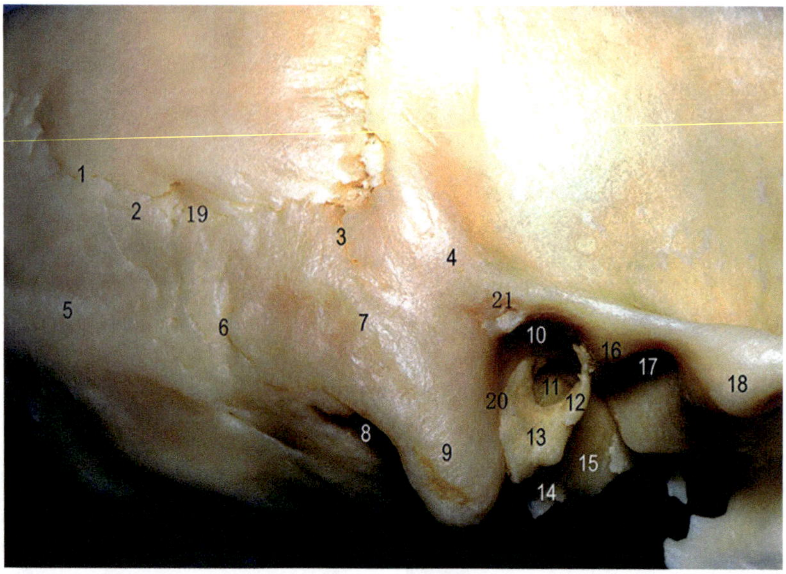

Fig. 3 Lateral view of right tympanic and mastoid part of temporal bone. (*1*) Lambdoid suture, (*2*) Superior nuchal line, (*3*) Squamosomastoid suture, (*4*) Suprameatal crest, (*5*) Inferior nuchal line, (*6*) Occipitomastoid suture, (*7*) Mastoid part of temporal bone, (*8*) Digastric groove, (*9*) Mastoid process, (*10*) Suprameatal triangle, (*11*) External acoustic meatus, (*12*) Anterior margin of external acoustic meatus, (*13*) Tympanic part of temporal bone, (*14*) Sheath of styloid process, (*15*) Pteryoid process, (*16*) Postglenoid process, (*17*) Mandibular fossa, (*18*) Articular tubercle, (*19*) Asterion, (*20*) Tympanomastiod fissure, (*21*) Suprameatal spine

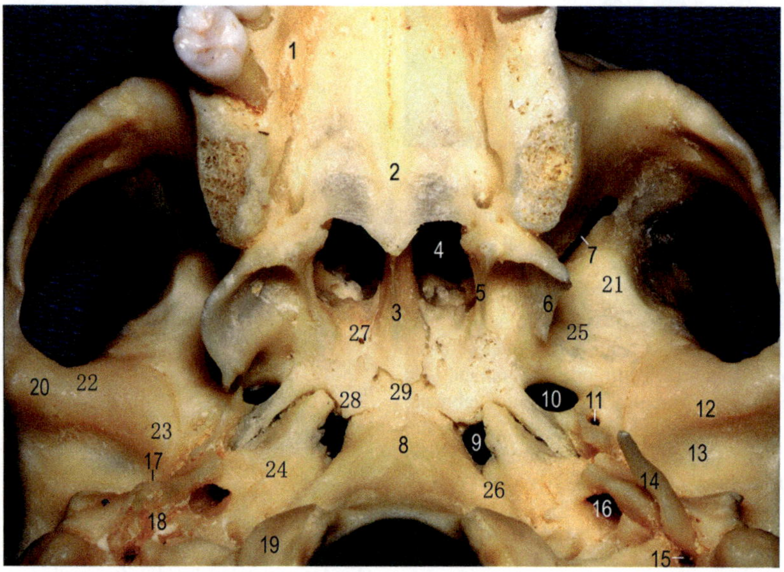

Fig. 4 External view of anterior and middle cranial base. (*1*) Maxilla, (*2*) Palatine bone, (*3*) Vomer, (*4*) Posterior nasal aperture, (*5*) Medial pterygoid plate, (*6*) Lateral pterygoid plate, (*7*) Inferior orbital fissure, (*8*) Clivus, (*9*) Foramen lacerum, (*10*) Foramen ovale, (*11*) Foramen spinosum, (*12*) Articular tubercle, (*13*) Mandibular fossa, (*14*) Styloid process, (*15*) Stylomastoid foramen, (*16*) External opening of carotid canal, (*17*) Margin of tegmen tympani, (*18*) Tympanic part of temporal bone, (*19*) Occipital condyle, (*20*) Zygomatic bone, (*21*) Greater wing of sphenoid bone, (*22*) Anterior root of zygomatic process, (*23*) Posterior root of zygomatic process, (*24*) Petrous part of temporal bone, (*25*) Infratemporal fossa, (*26*) Petroclival fissure, (*27*) Sphenoidal conchae, (*28*) Lingual process of sphenoidal bone, (*29*) Sphenoidal rostrum

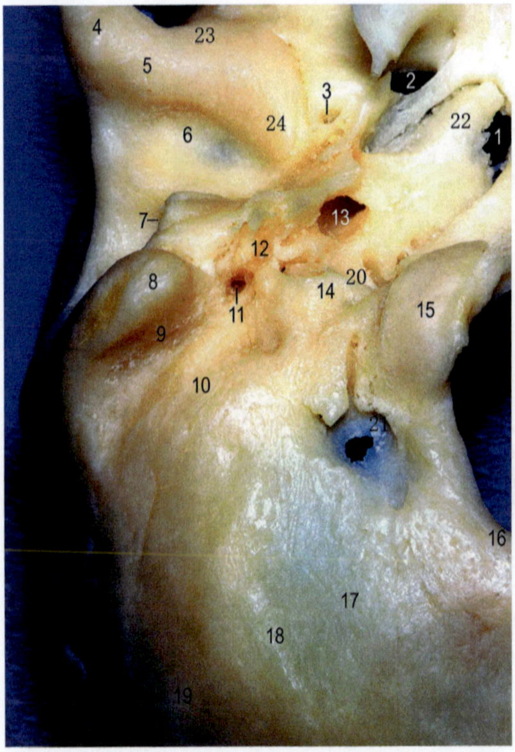

Fig. 5 Inferior view of right middle and posterior cranial fossae. (*1*) Foramen lacerum, (*2*) Foramen ovale, (*3*) Foramen spinosum, (*4*) Root of zygomatic arch, (*5*) Articular tubercle, (*6*) Mandibular fossa, (*7*) External acoustic meatus, (*8*) Mastoid process, (*9*) Digastric groove, (*10*) Occipital groove, (*11*) Stylomastoid foramen, (*12*) Tympanic part of temporal bone, (*13*) External opening of carotid canal, (*14*) Jugular process, (*15*) Occipital condyle, (*16*) Posterior margin of foramen magnum, (*17*) Squamous part of occipital bone, (*18*) Inferior nuchal line, (*19*) Superior nuchal line, (*20*) Jugular foramen, (*21*) Posterior condylar canal, (*22*) Petrous apex, (*23*) Anterior root of zygomatic process, (*24*) Posterior root of zygomatic process

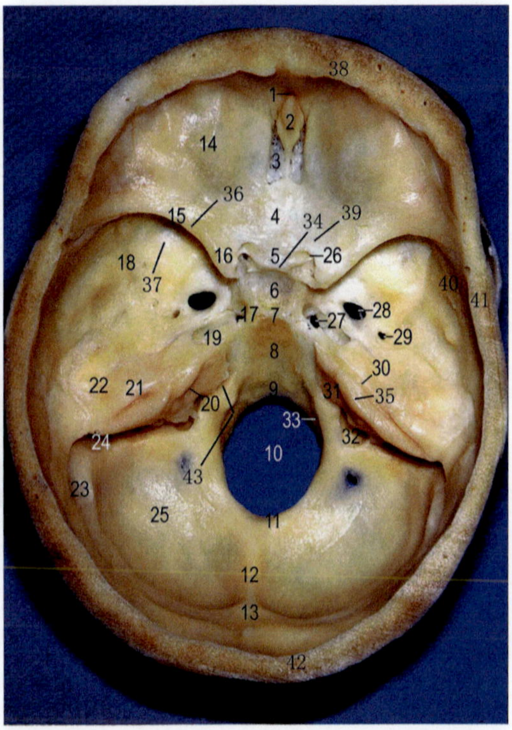

Fig. 6 Internal view of the skull base. (*1*) Foramen cecum, (*2*) Crista galli, (*3*) Cribriform plate, (*4*) Planum sphenoidale, (*5*) Chiasmatic sulcus, (*6*) Pituitary fossa, (*7*) Dorsum sellae, (*8*) Clivus, (*9*) Anterior margin of foramen magnum, (*10*) Foramen magnum, (*11*) Posterior margin of foramen magnum, (*12*) Internal occipital crest, (*13*) Internal occipital protuberance, (*14*) Anterior cranial fossa, (*15*) Sphenoid ridge, (*16*) Anterior clinoid process, (*17*) Posterior clinoid process, (*18*) Middle cranial fossa, (*19*) Petrous apex, (*20*) Internal acoustic meatus, (*21*) Arcuate eminence, (*22*) Tegmen tympani, (*23*) Transverse sulcus, (*24*) Sigmoid sulcus, (*25*) Posterior cranial fossa, (*26*) Optic canal, (*27*) Foramen lacerum, (*28*) Foramen ovale, (*29*) Foramen spinosum, (*30*) Superior petrosal sulcus, (*31*) Inferior petrosal sulcus, (*32*) Jugular foramen, (*33*) Hypoglossal canal, (*34*) Tuberculum sellae, (*35*) Petrous ridge, (*36*) Lesser wing of sphenoid bone, (*37*) Greater wing of sphenoid bone, (*38*) Frontal bone, (*39*) Sphenoid bone, (*40*) Temporal bone, (*41*) Parietal bone, (*42*) Occipital bone, (*43*) Petroclival fissure

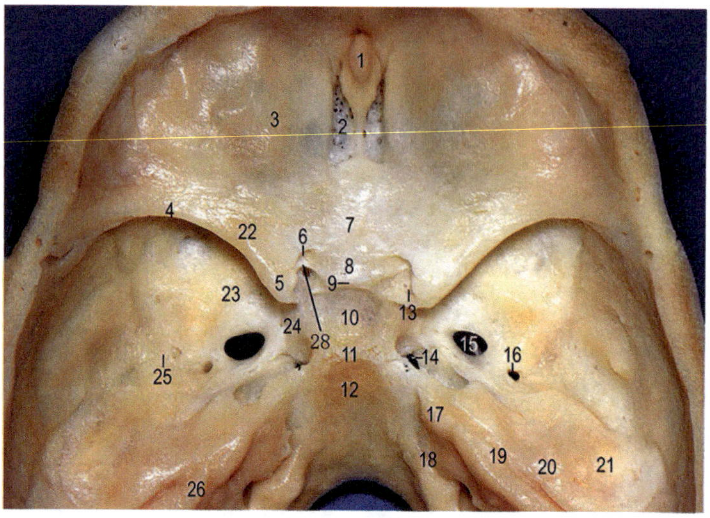

Fig. 7 Internal view of anterior and middle cranial fossae. (*1*) Crista galli, (*2*) Cribriform plate, (*3*) Anterior cranial fossa, (*4*) Sphenoid ridge, (*5*) Anterior clinoid process, (*6*) Optic canal, (*7*) Planum sphenoidale, (*8*) Chiasmatic sulcus, (*9*) Tuberculum sellae, (*10*) Pituitary fossa, (*11*) Dorsum sellae, (*12*) Clivus, (*13*) Caroticoclinoid foramen, (*14*) Internal opening of carotid canal, (*15*) Foramen ovale, (*16*) Foramen spinosum, (*17*) Petrous apex, (*18*) Inferior petrosal sulcus, (*19*) Superior petrosal sulcus, (*20*) Arcuate eminence, (*21*) Tegmen tympani, (*22*) Lesser wing of sphenoid bone, (*23*) Greater wing of sphenoid bone, (*24*) Carotid sulcus, (*25*) Groove for middle meningeal artery, (*26*) Petrous ridge, (*27*) NA, (*28*) Optic strut

Fig. 8 Internal view of middle and posterior cranial fossae. (*1*) Tuberculum sellae, (*2*) Pituitary fossa, (*3*) Dorsum sellae, (*4*) Anterior clinoid process, (*5*) Foramen rotundum, (*6*) Internal opening of carotid canal, (*7*) Foramen ovale, (*8*) Petrous apex, (*9*) Superior petrosal sulcus, (*10*) Arcuate eminence, (*11*) Tegmen tympani, (*12*) Dorsum sellae, (*13*) Inferior petrosal sulcus, (*14*) Internal acoustic meatus, (*15*) Jugular foramen, (*16*) Posterior jugular ridge, (*17*) Sigmoid sulcus, (*18*) Transverse sulcus, (*19*) Petrous ridge, (*20*) Middle clinoid process, (*21*) Caroticoclinoid foramen, (*22*) Carotid sulcus, (*23*) Groove for middle meningeal artery, (*24*) Trigeminal impression, (*25*) Endolymphatic depression

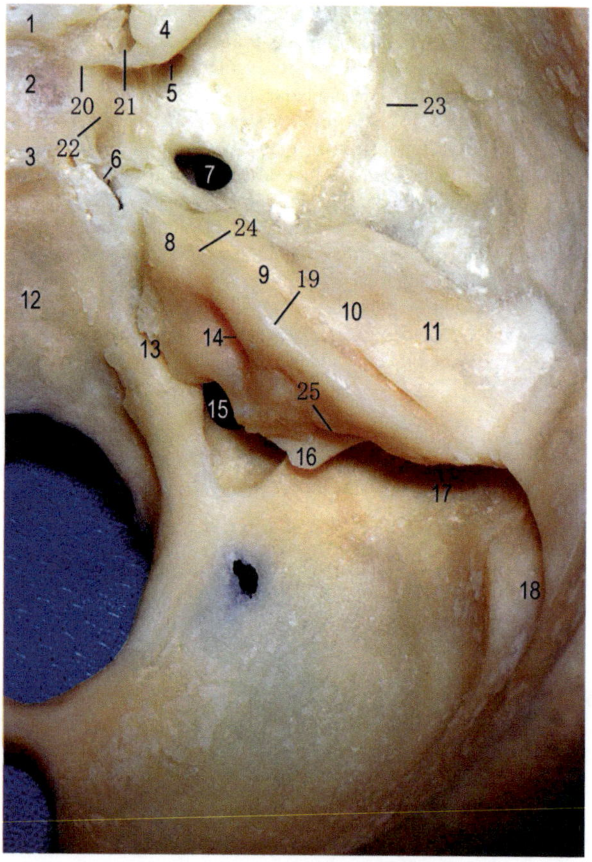

2 Anatomy of Dura Mater

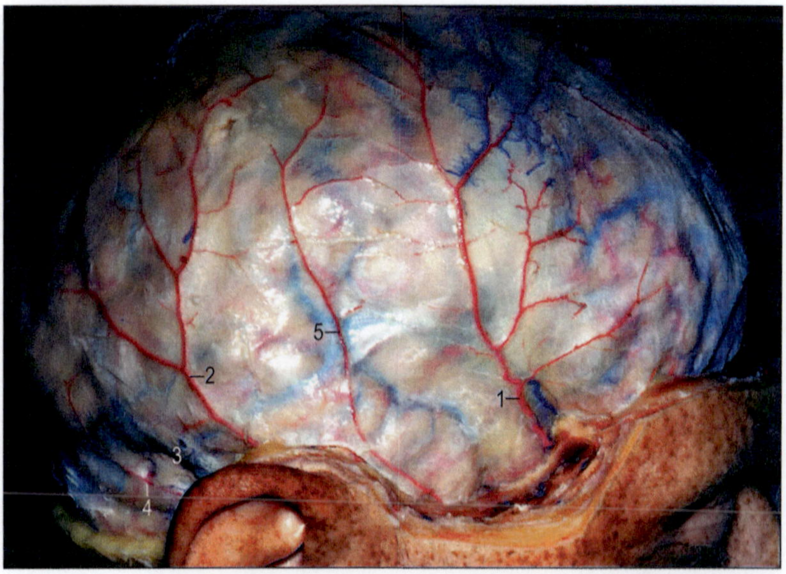

Fig. 9 Lateral view of cranial dura mater. (*1*) Frontal branch of middle meningeal artery, (*2*) Parietooccipital branch of middle meningeal artery, (*3*) Transverse sinus, (*4*) Posterior meningeal artery, (*5*) Parietal branch of middle meningeal artery

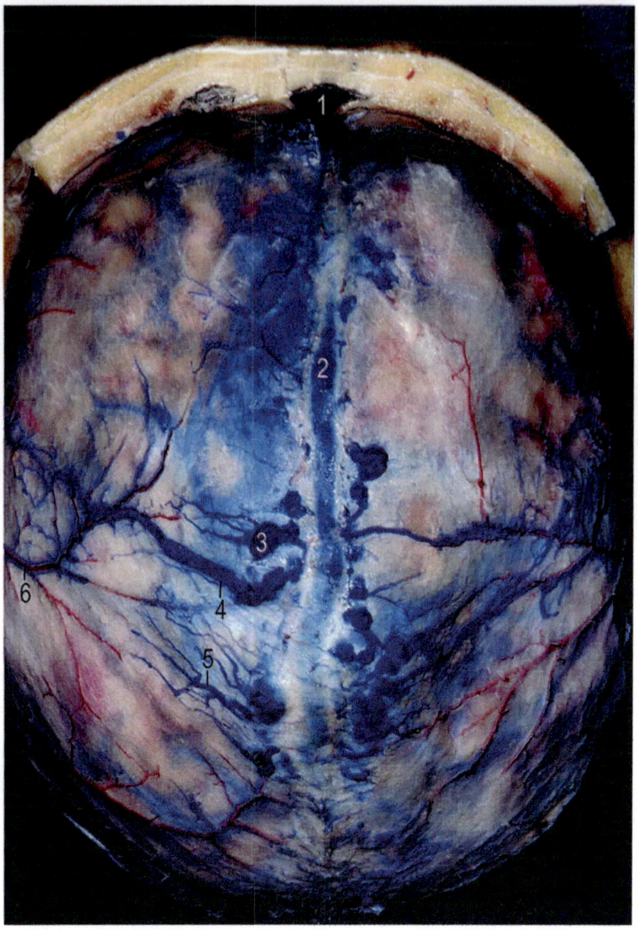

Fig. 10 Superior view of convexity dura mater. (*1*) Frontal sinus, (*2*) Superior sagittal sinus, (*3*) Lateral lacunae, (*4*) Middle meningeal veins, (*5*) Posterior parietal meningeal vein, (*6*) Middle meningeal artery

3 Anatomy of Brain Surface

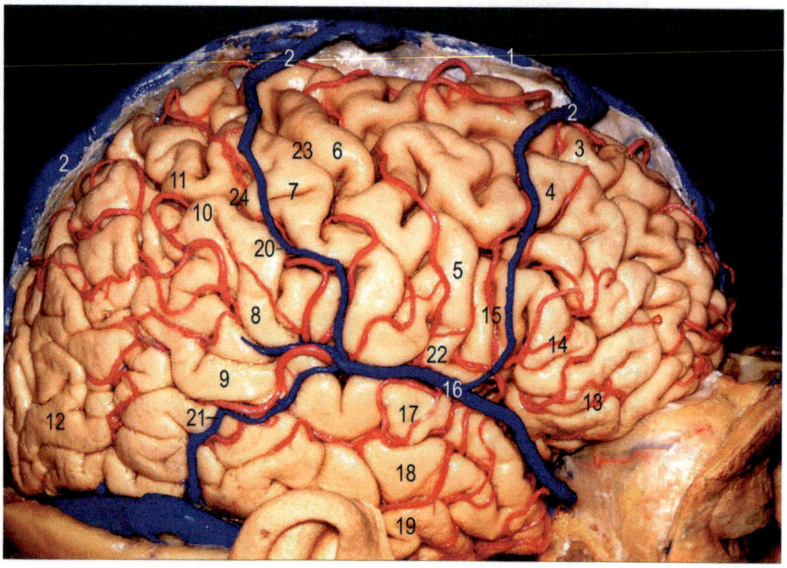

Fig. 11 Overview of lateral surface of the brain. (*1*) Superior sagittal sinus, (*2*) Superior cerebral veins, (*3*) Superior frontal gyrus, (*4*) Middle frontal gyrus, (*5*) Inferior frontal gyrus, (*6*) Precentral gyrus, (*7*) Postcentral gyrus, (*8*) Supramarginal gyrus, (*9*) Angular gyrus, (*10*) Inferior parietal lobule, (*11*) Superior parietal lobule, (*12*) Occipital lobe, (*13*) Pars Orbitalis, (*14*) Pars Triangularis, (*15*) Anterior part of Pars opercularis, (*16*) Sylvian vein (superficial middle cerebral vein), (*17*) Superior temporal gyrus, (*18*) Middle temporal gyrus, (*19*) Inferior temporal gyrus, (*20*) Superior anastomotic vein (vein of Troland), (*21*) Inferior anastomotic vein (vein of Labbe), (*22*) Posterior part of Pars opercularis, (*23*) Central sulcus, (*24*) Postcentral sulcus

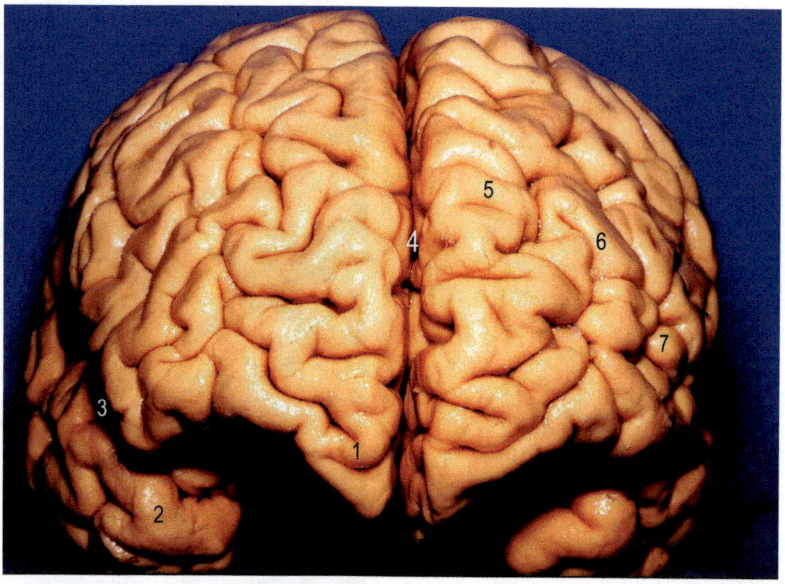

Fig. 12 Anterior view of lateral surface of the brain. (*1*) Frontal pole, (*2*) Temporal pole, (*3*) Sylvian fissure, (*4*) Longitudinal fissure, (*5*) Superior frontal gyrus, (*6*) Middle frontal gyrus, (*7*) Inferior frontal gyrus

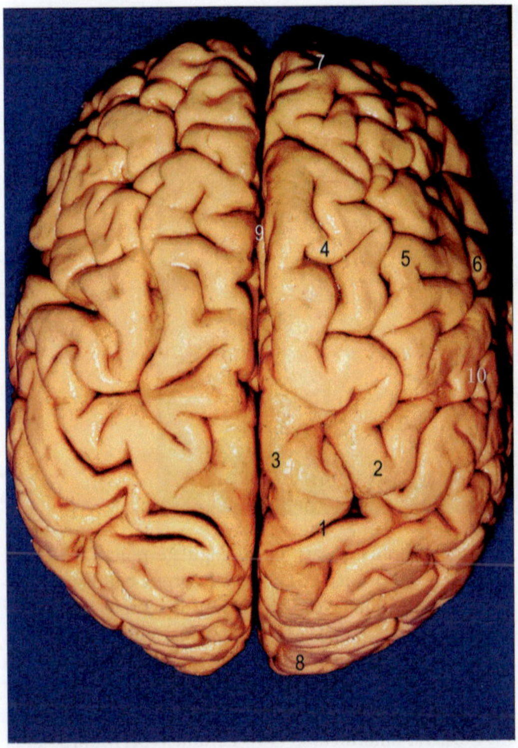

Fig. 13 Superior view of the lateral surface of the brain. (*1*) Central sulcus, (*2*) Precentral gyrus, (*3*) Paracentral lobule, (*4*) Superior frontal sulcus, (*5*) Pars triangularis, (*6*) Inferior frontal gyrus, (*7*) Frontal pole, (*8*) Occipital pole, (*9*) Longitudinal fissure, (*10*) Pars opercularis

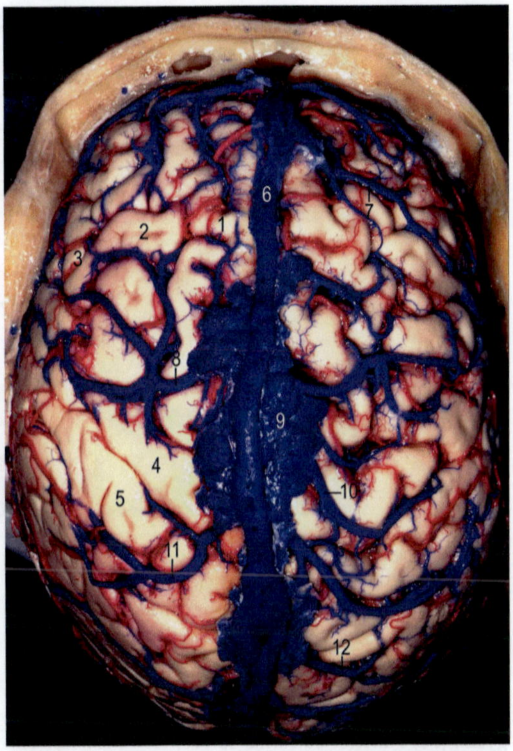

Fig. 14 Superior sagittal sinus and its tributaries. (*1*) Superior frontal gyrus, (*2*) Middle frontal gyrus, (*3*) Inferior frontal gyrus, (*4*) Precentral gyrus, (*5*) Postcentral gyrus, (*6*) Frontal part of superior sagittal sinus, (*7*) Middle frontal vein, (*8*) Posterior frontal vein, (*9*) Venous Lacunae, (*10*) Central vein, (*11*) Postcentral vein, (*12*) Anterior parietal vein

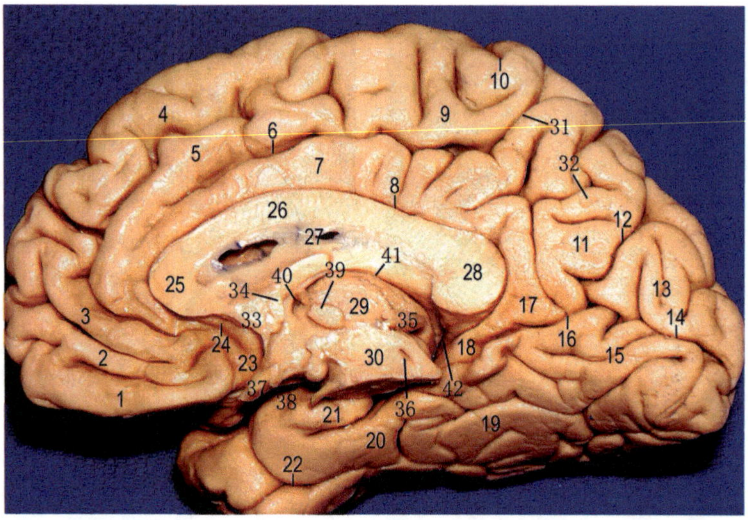

Fig. 15 Overview of medial surface of the brain. (*1*) Gyrus rectus, (*2*) Inferior rostral gyrus, (*3*) Superior rostral gyrus, (*4*) Medial superior frontal gyrus, (*5*) Paracingulate gyrus, (*6*) Cingulate sulcus, (*7*) Cingulate gyrus, (*8*) Callosal sulcus, (*9*) Paracentral lobule, (*10*) Central sulcus, (*11*) Precuneus, (*12*) Parieto-occipital sulcus, (*13*) Cuneus, (*14*) Calcarine sulcus, (*15*) Lingual gyrus, (*16*) Anterior calcarine sulcus, (*17*) Posterior cingulate gyrus, (*18*) Isthmus, (*19*) Fusiform gyrus, (*20*) Parahippocampal gyrus, (*21*) Uncus, (*22*) Rhinal sulcus, (*23*) Paraterminal gyrus, (*24*) Cingulate pole, (*25*) Genu of corpus callosum, (*26*) Body of corpus callosum, (*27*) Septum pellucidum, (*28*) Splenium of corpus callosum, (*29*) Thalamus, (*30*) Mesencephalon, (*31*) Ascending ramus of cingulate sulcus, (*32*) Precuneus, (*33*) Anterior commissure, (*34*) Column of fornix, (*35*) Pineal body, (*36*) Cerebral aqueduct, (*37*) Optic chiasm, (*38*) Apex, (*39*) Massa intermedia, (*40*) Foramen of Monro, (*41*) Choroid fissure, (*42*) Fasciolar gyrus

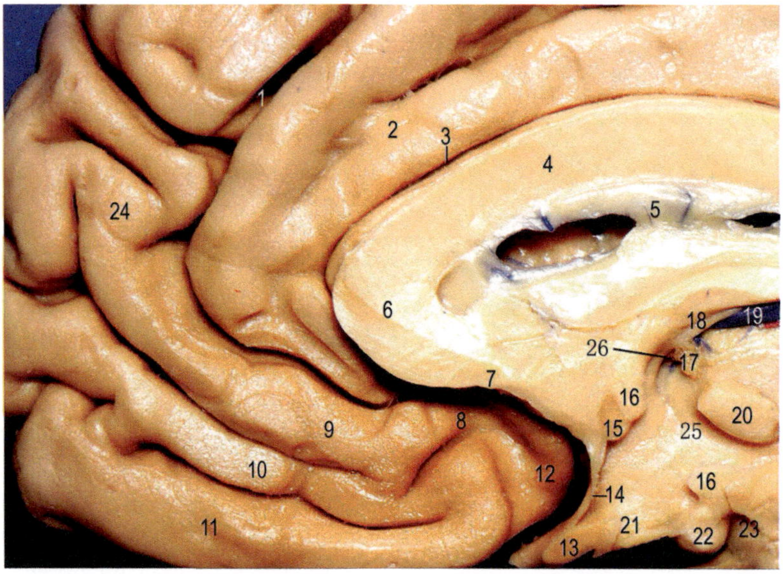

Fig. 16 Sagittal view of anterior medial frontal region. (*1*) Cingulate sulcus, (*2*) Cingulate gyrus, (*3*) Callosal sulcus, (*4*) Body of corpus callosum, (*5*) Septum pellucidum, (*6*) Genu of corpus callosum, (*7*) Rostrum of corpus callosum, (*8*) Cingulate Pole, (*9*) Superior rostral gyrus, (*10*) Inferior rostral gyrus, (*11*) Rectus gyrus, (*12*) Postolfactory sulcus, (*13*) Optic chiasm, (*14*) Lamina terminalis, (*15*) Anterior commissure, (*16*) Column of fornix, (*17*) Choroid plexus, (*18*) Body of fornix, (*19*) Internal cerebral vein, (*20*) Massa intermedia, (*21*) Infundibular recess, (*22*) Mamillary body, (*23*) Interpeduncular fossa, (*24*) Medial frontal gyrus, (*25*) Hypothalamic sulcus, (*26*) Foramen of Monro

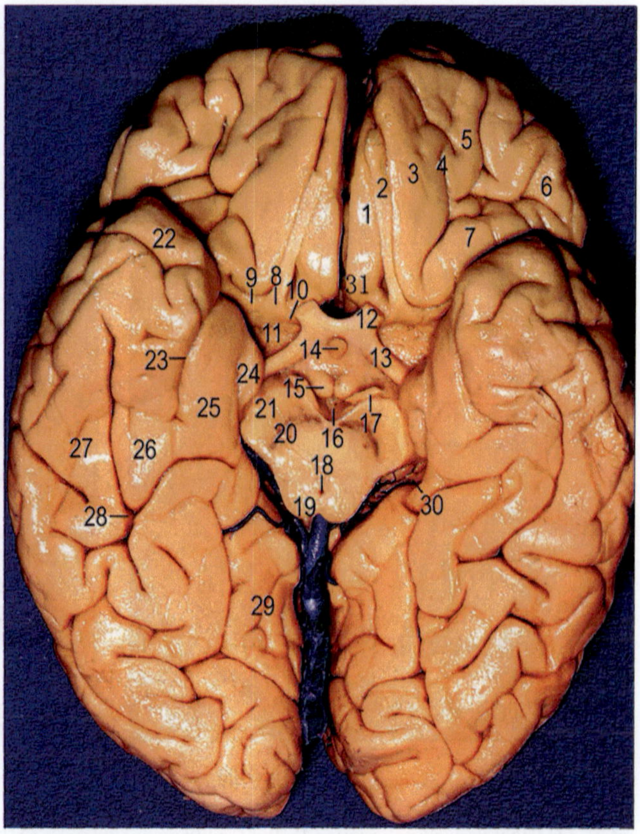

Fig. 17 Overview of inferior surface of the brain. (*1*) Gyrus rectus, (*2*) Olfactory tract, (*3*) Medial orbital gyrus, (*4*) Orbital sulcus, (*5*) Anterior orbital gyrus, (*6*) Lateral orbital gyrus, (*7*) Posterior orbital gyrus, (*8*) Olfactory trigone, (*9*) Lateral olfactory stria, (*10*) Medial olfactory stria, (*11*) Anterior perforator substance, (*12*) Optic nerve, (*13*) Optic tract, (*14*) Infundibulum, (*15*) Mammillary body, (*16*) Posterior perforator substance, (*17*) Oculomotor nerve, (*18*) Cerebral aqueduct, (*19*) Superior colliculus, (*20*) Substantia nigra, (*21*) Cerebral peduncle, (*22*) Temporal pole, (*23*) Rhinal sulcus, (*24*) Uncus, (*25*) Parahippocampal gyrus, (*26*) Fusiform gyrus, (*27*) Inferior temporal gyrus, (*28*) Occipitotemporal sulcus, (*29*) Lingual gyrus, (*30*) Lateral geniculate body, (*31*) Paraterminal gyrus

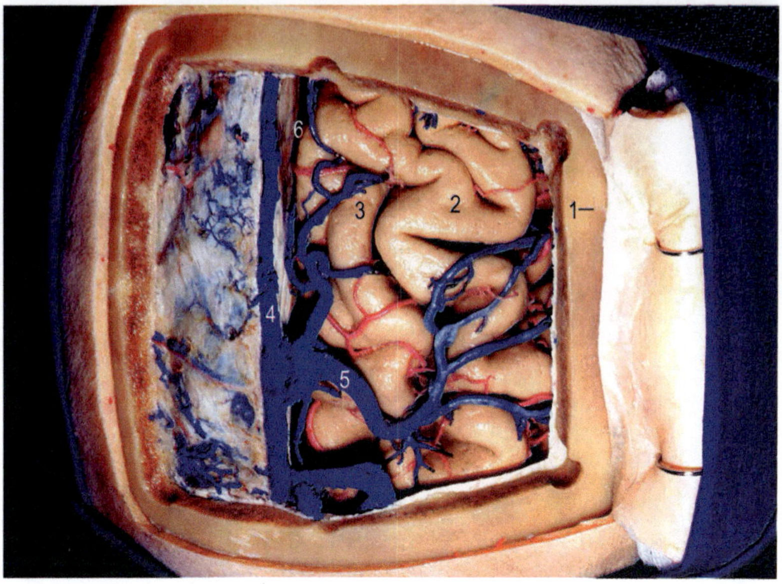

Fig. 18 Craniotomy of interhemispheric transcallosal approach. (*1*) Coronal suture, (*2*) Middle frontal gyrus, (*3*) Superior frontal gyrus, (*4*) Superior sagittal sinus, (*5*) Posterior frontal vein, (*6*) Longitudinal fissure

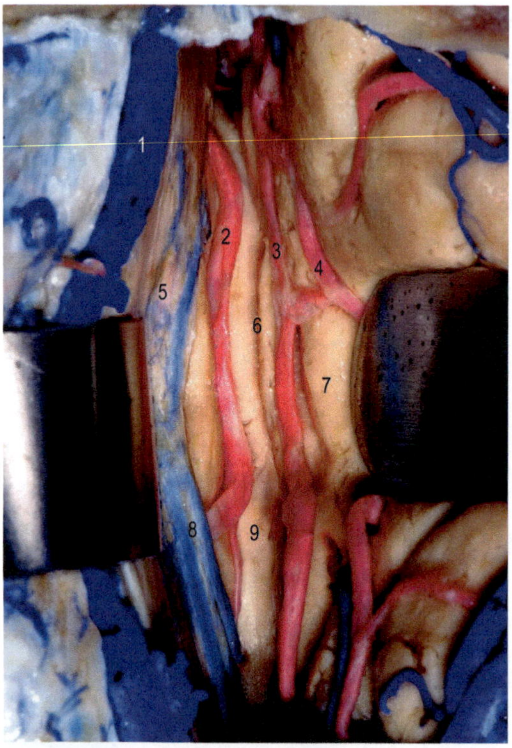

Fig. 19 Trajectory of interhemispheric transcallosal approach. (*1*) Superior sagittal sinus, (*2*) Left pericallosal artery, (*3*) Right pericallosal artery, (*4*) Right callosomarginal artery, (*5*) Falx cerebri, (*6*) Right callosal sulcus, (*7*) Cingulate gyrus, (*8*) Inferior sagittal sinus, (*9*) Body of corpus callosum

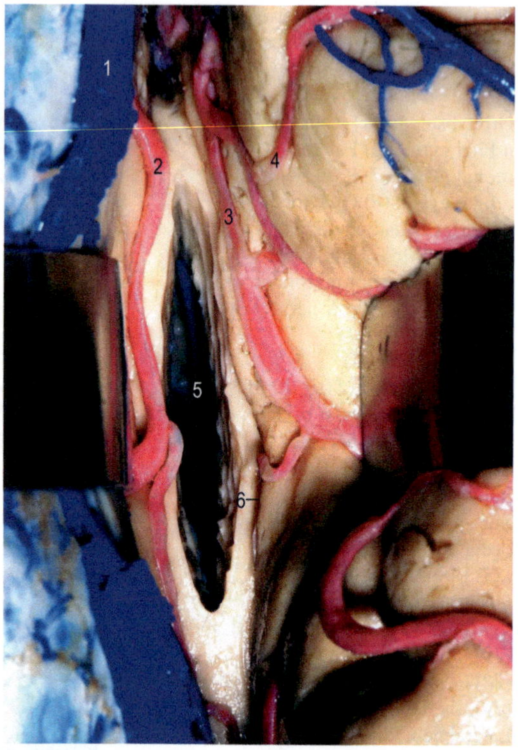

Fig. 20 Callostomy of interhemispheric transcallosal approach, (*1*) Superior sagittal sinus, (*2*) Left pericallosal artery, (*3*) Right pericallosal artery, (*4*) Right medial intermediate frontal artery, (*5*) Third ventricle, (*6*) Right callosal sulcus

Case 1.1 Frontal Epidermoid Cyst

Clinical presentation:

A 12-year-old boy presented with a 3-year history of frontal mass.

Neurological exam:

Uneventful

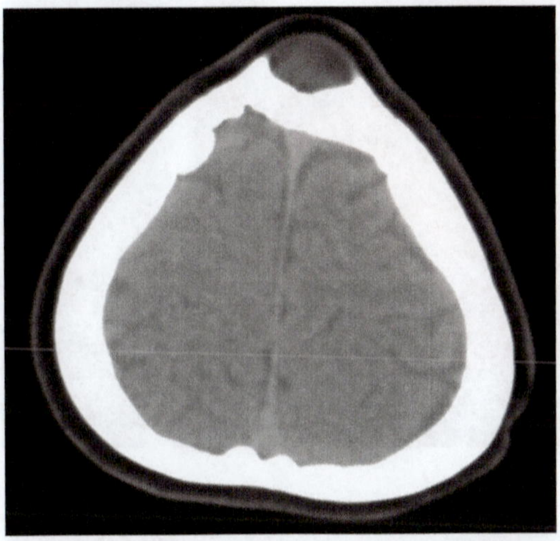

Fig. 21 Pre-op CT scan revealed a well-defined, uncalcified, and hypodense mass eroding the outer table of frontal bone

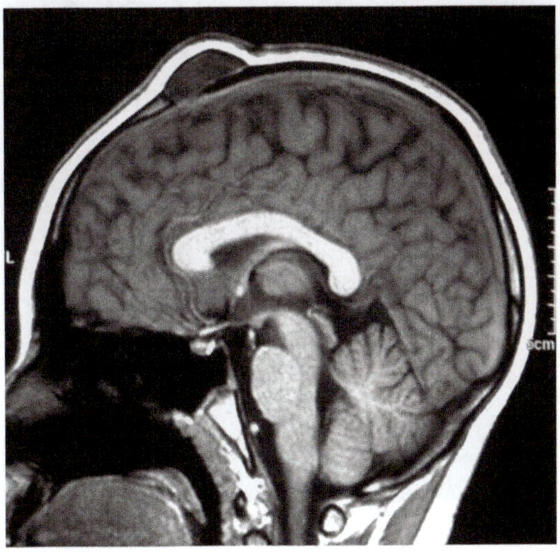

Fig. 22 Sagittal T1-weighted MRI

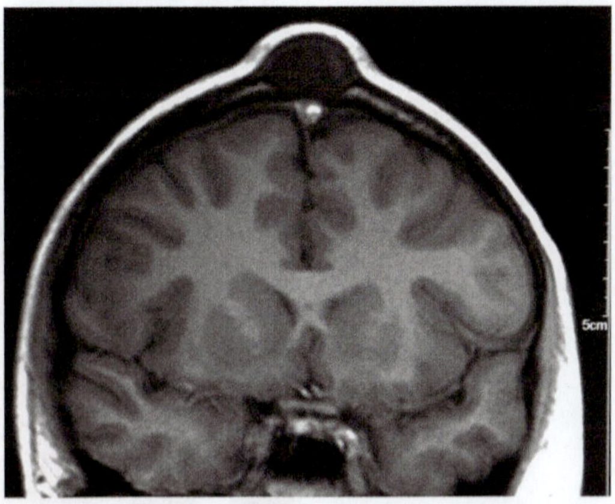

Fig. 23 Coronal T1-weighted MRI. Pre-op T1-weighted MRI showed an osseous round lesion expanding into subcutaneous tissue and invading the inner table of frontal bone

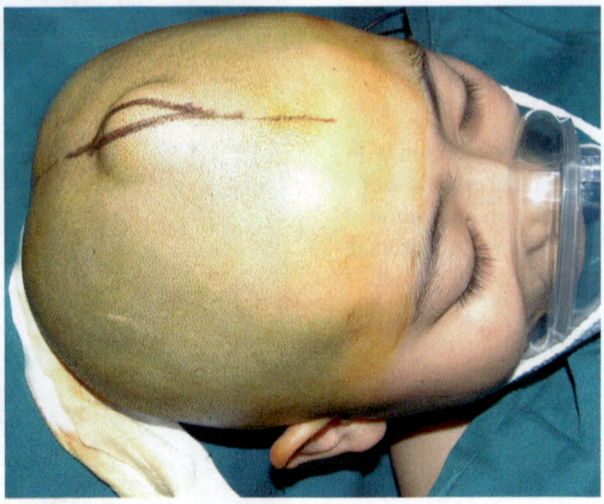

Fig. 24 A curvilinear incision was made along the outer margin of osseous mass

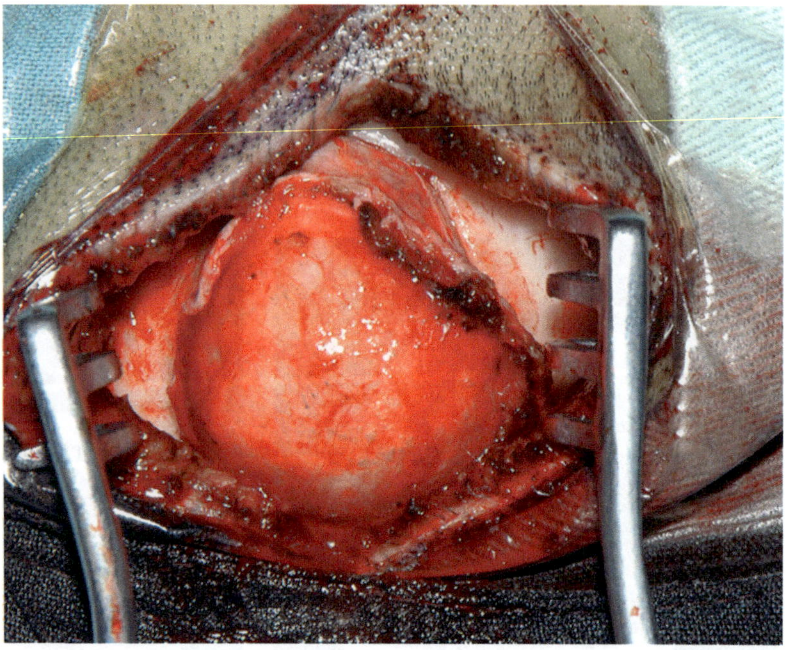

Fig. 25 The nodule protruded from the bony surface with ductile texture after opening the subcutaneous tissue

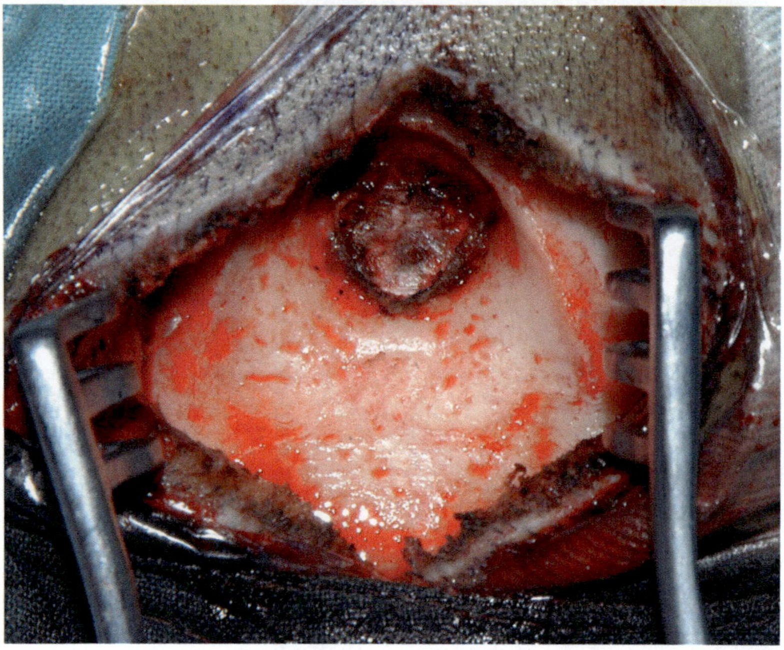

Fig. 26 The lesion with outer table erosion was resected en bloc while the inner table of frontal bone was well defined

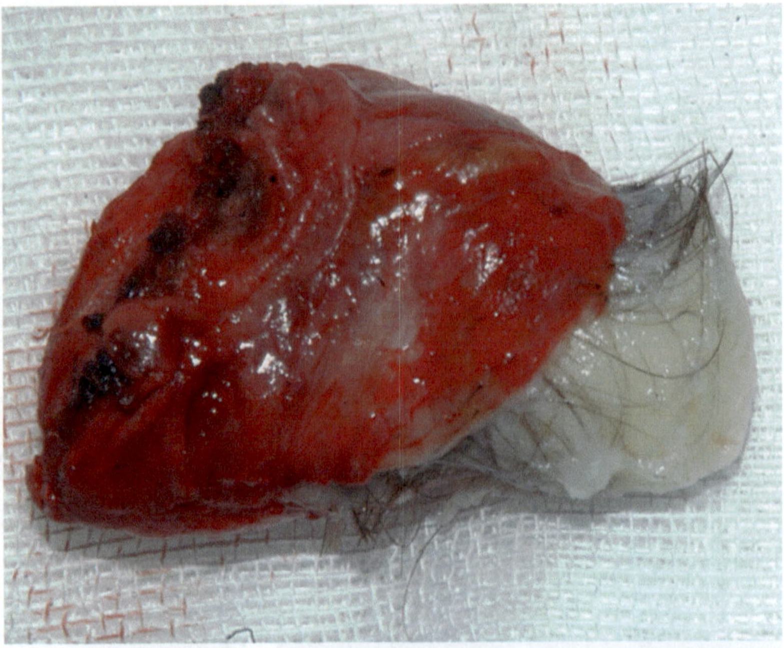

Fig. 27 The mass was opened and the lipid component was seen with hair

Case 1.2 Right Frontoparietal Meningioma

Clinical presentation:

A 39-year-old lady presented with a history of headache and hemiparesis of right lower extremity for 4 days.

Neurological exam:

Muscle strength: Grade IV (Lt lower extremity), others V

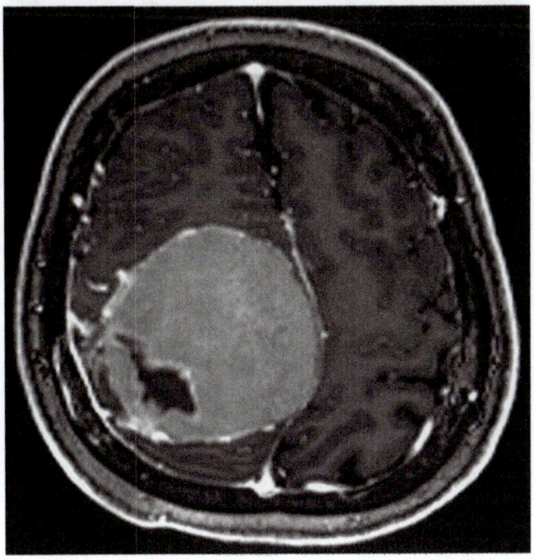

Fig. 28 Pre-op axial Gd-enhanced T1-weighted MR

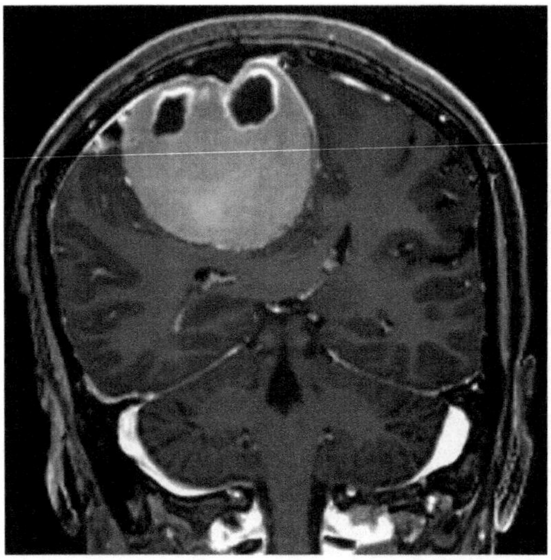

Fig. 29 Pre-op coronal Gd-enhanced T1-weighted MRI

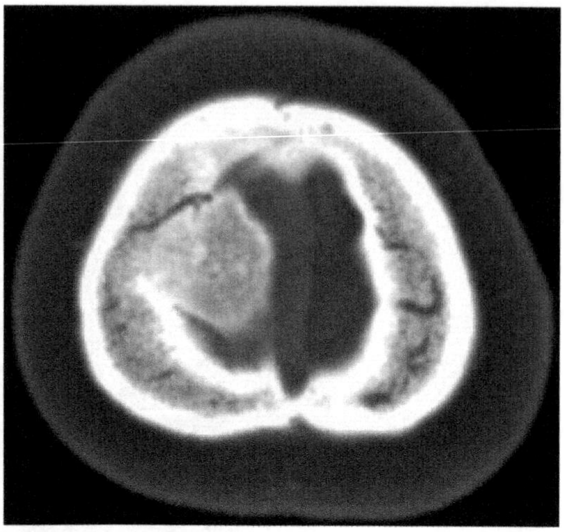

Fig. 31 Pre-op axial CT (bone window)

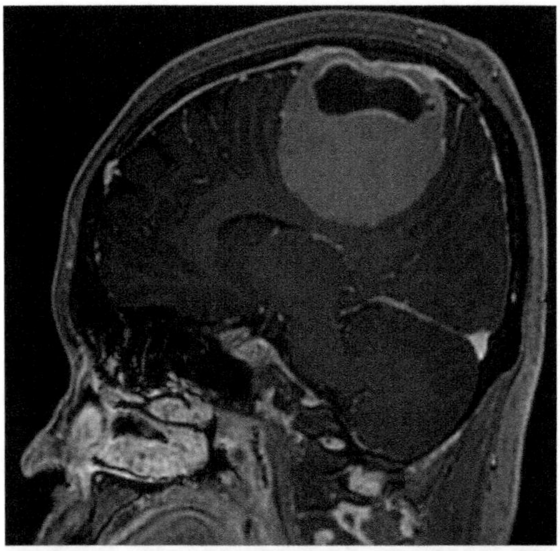

Fig. 30 Pre-op sagittal Gd-enhanced T1-weighted MRI. Pre-op Gd-enhanced T1-weighted MRI delineated a well-circumscribed frontoparietal lesion that was enhanced heterogeneously. The mass effect has been created pushing the falx into contralateral side and dura tail was seen

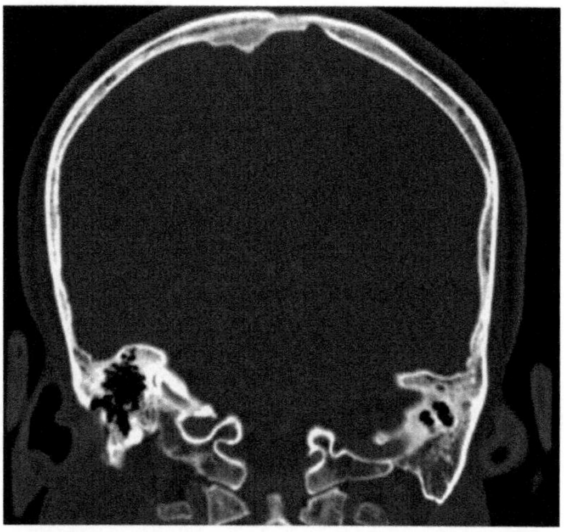

Fig. 32 Pre-op coronal CT (bone window)

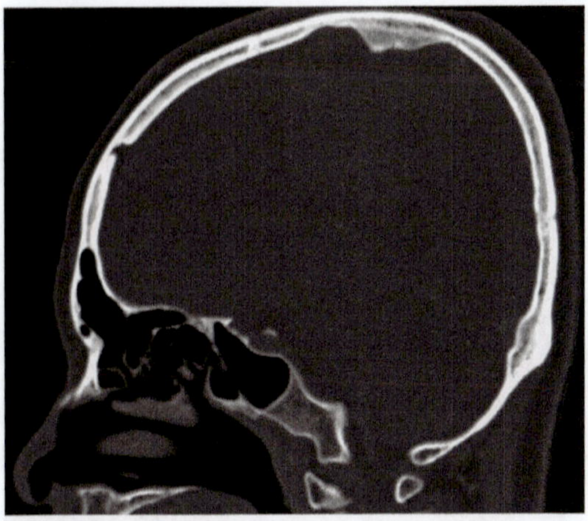

Fig. 33 Pre-op sagittal CT (bone window)

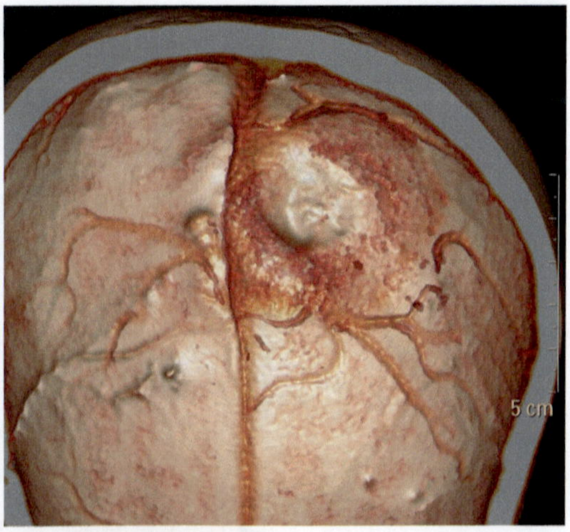

Fig. 34 Pre-op 3D CT. Pre-op CT scan with bone window illustrated hyperostosis of overlaying calvarium

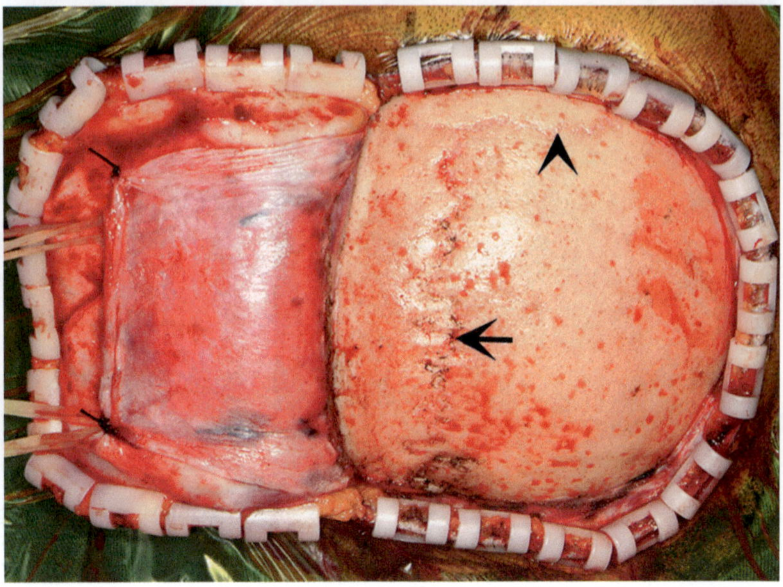

Fig. 35 Supine position was carried out and skin flap was retracted to the midline. Sagittal (arrow) and coronal suture (arrowhead) was exposed

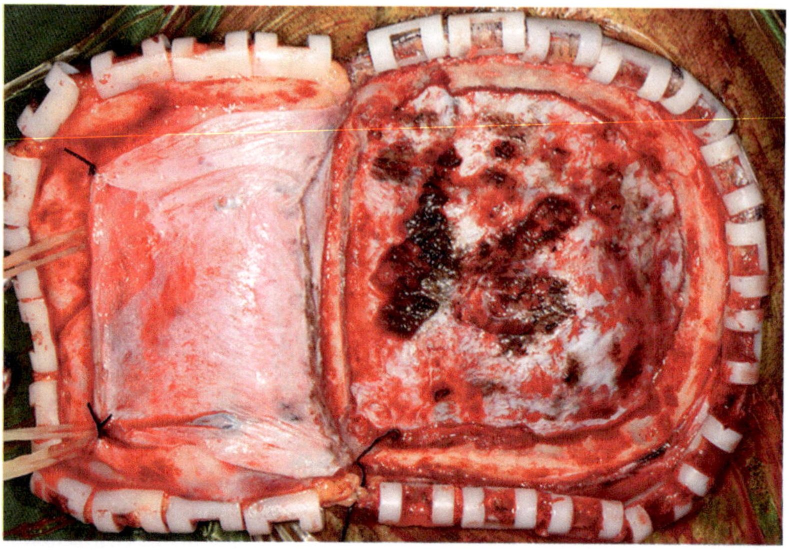

Fig. 36 The frontoparietal dura has been eroded by the tumor

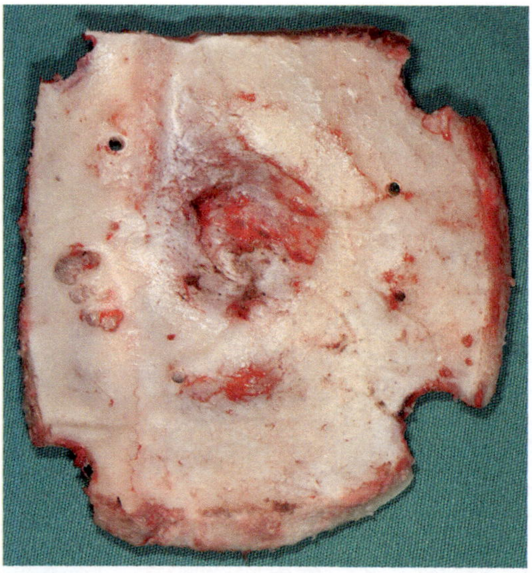

Fig. 37 The inner plate of overlaying calvarium was also got involved in inducing hyperostosis

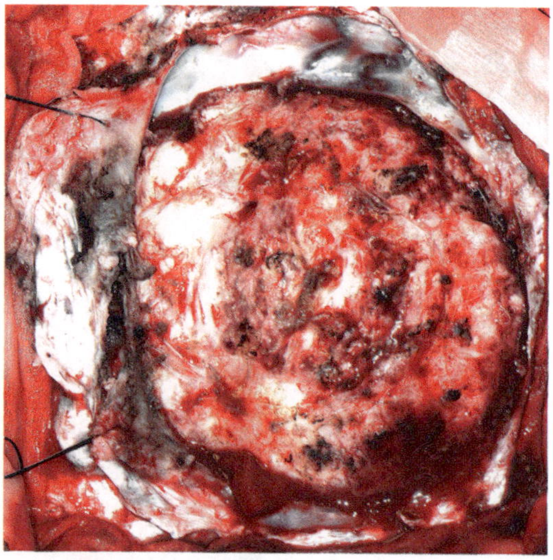

Fig. 38 After dura opening, the giant mass was unveiled and a well-defined margin was created to release the tumor from surrounding parenchyma

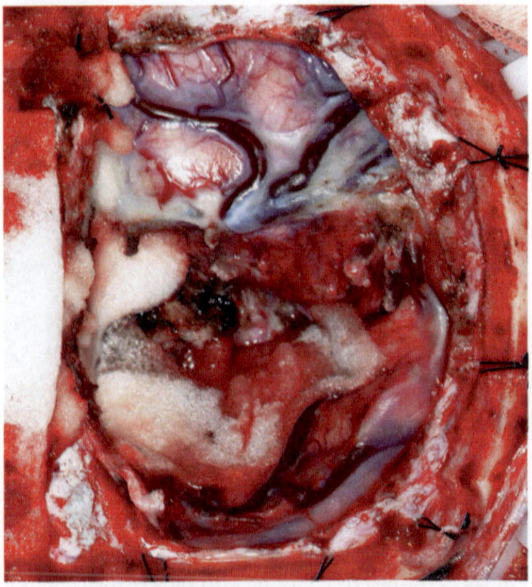

Fig. 39 Tumor was removed in a piecemeal fashion and perinidual brain tissue was compressed with gel foam for definite hemostasis

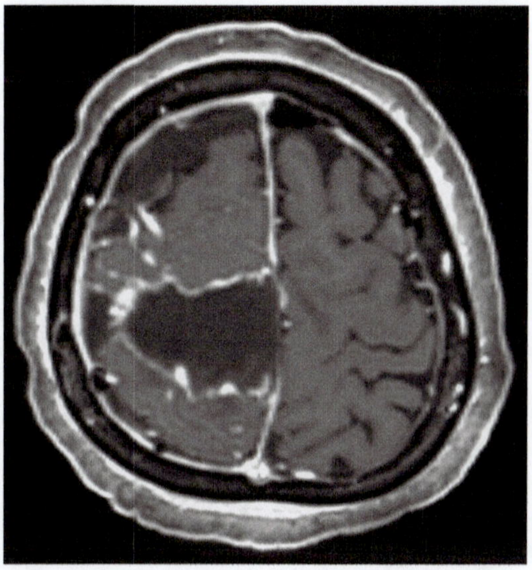

Fig. 41 Post-op axial Gd-enhanced T1-weighted MRI

Fig. 40 Closure was performed in a standard fashion

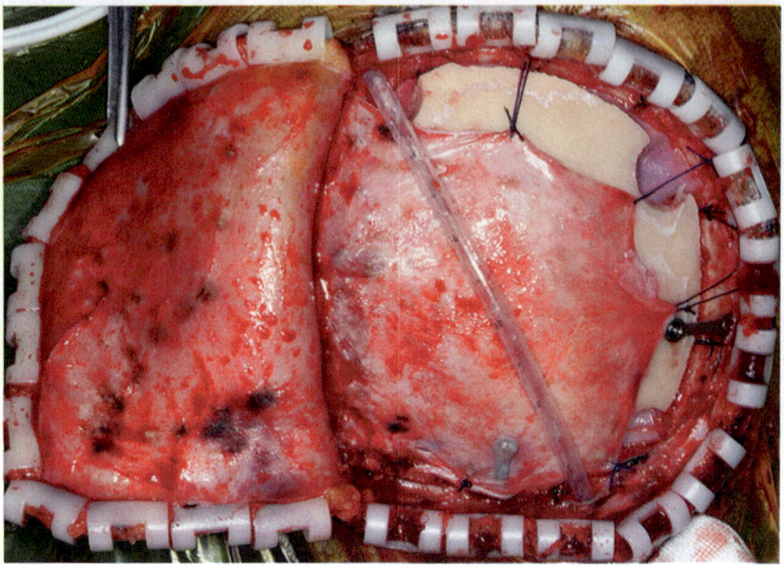

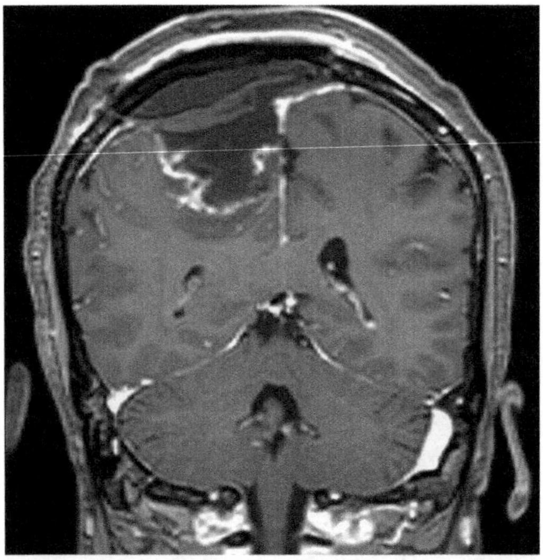

Fig. 42 Post-op coronal Gd-enhanced T1-weighted MRI

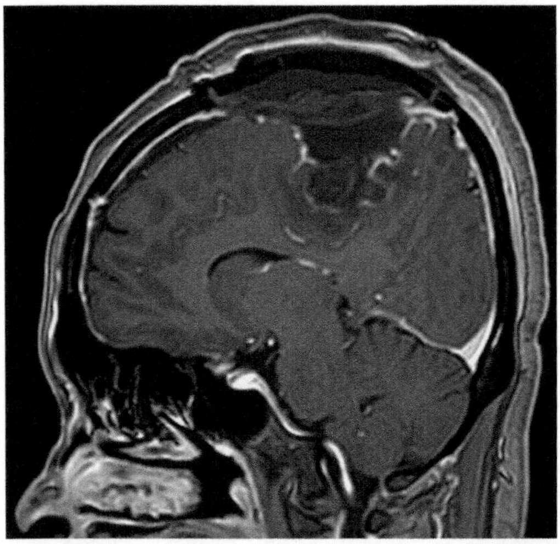

Fig. 43 Post-op sagittal Gd-enhanced T1-weighted MRI

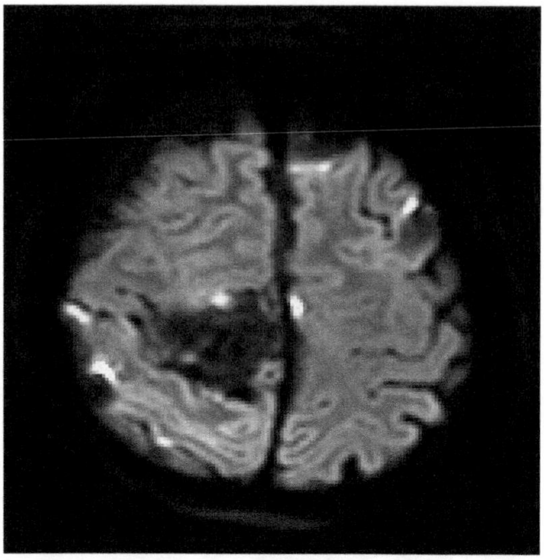

Fig. 44 Post-op DWI. Post-op MRI showed total removal of the mass without perinidual cortical infraction. Her muscle strength of left lower extremity was stable as pre-op baseline

Case 1.3 Falx Meningioma

Clinical presentation:

A 56-year-old female presented with a 7-year history of headache and exacerbated within last 10 days.

Neurological exam:

Unremarkable

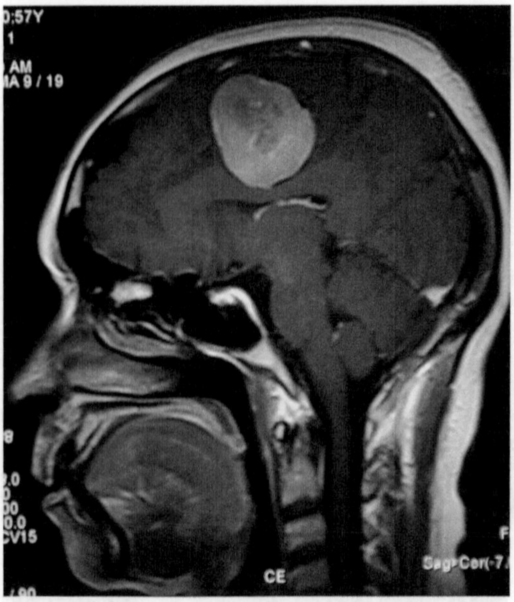

Fig. 45 Pre-op sagittal T1 contrast-enhanced MRI

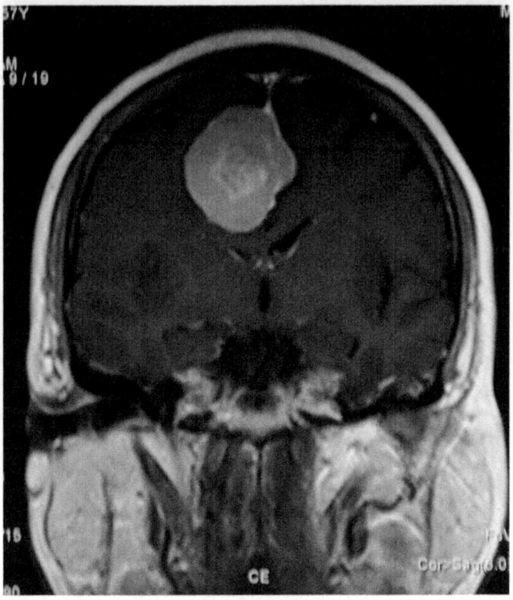

Fig. 46 Pre-op coronal T1 contrast-enhanced MRI

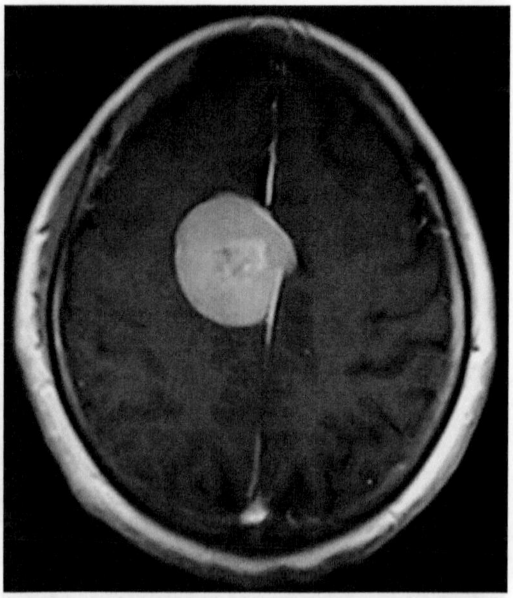

Fig. 47 Pre-op axial T1 contrast-enhanced MRI. Pre-op T1 contrast-enhanced MRI demonstrated a right frontal mass with involvement of anterior one-third falx. This well-defined tumor was homogeneously enhanced and dural tail sign was mentioned

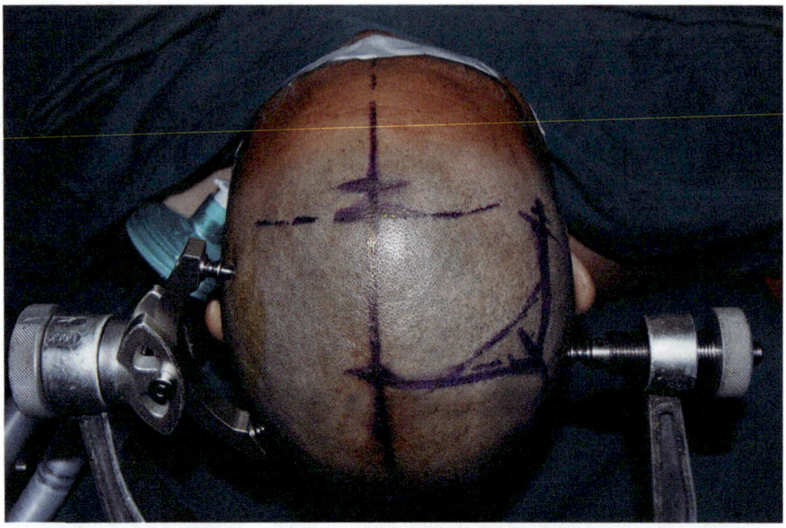

Fig. 48 Supine position was place and right frontoparietal incision was marked to reflect skin flap anteriorly

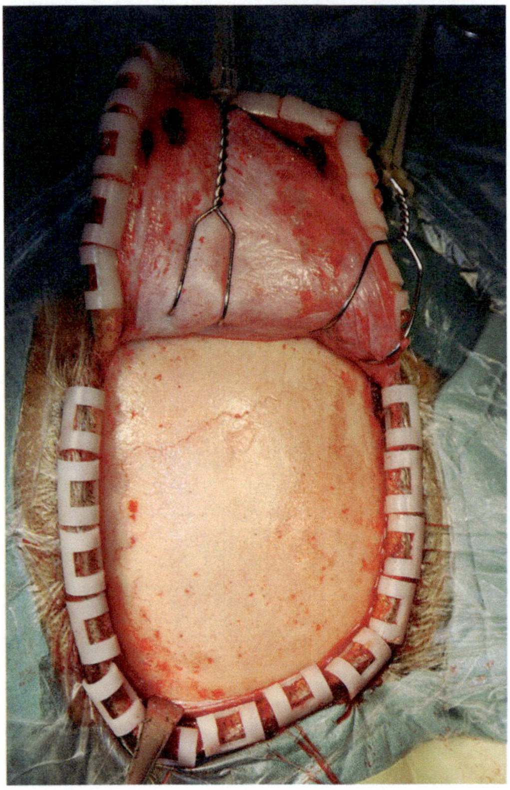

Fig. 49 Sagittal suture, coronal suture and intersection point, Bregma, can be visualized

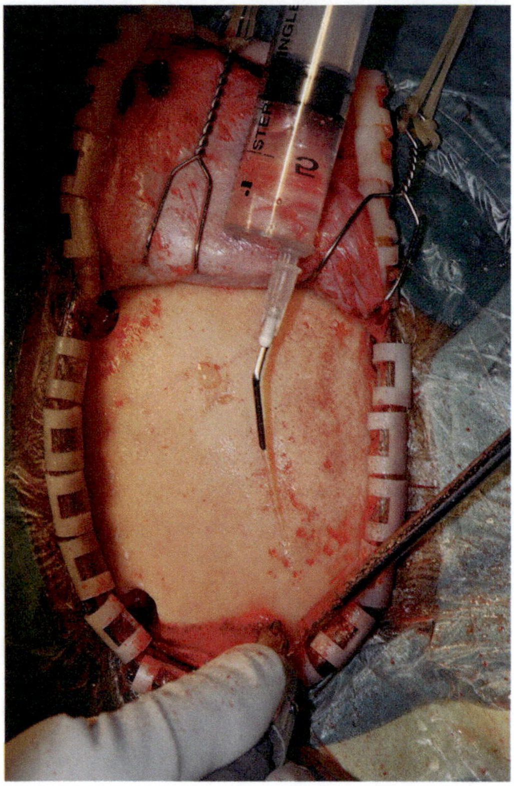

Fig. 50 Two burr holes were created at midline/superior sagittal sinus

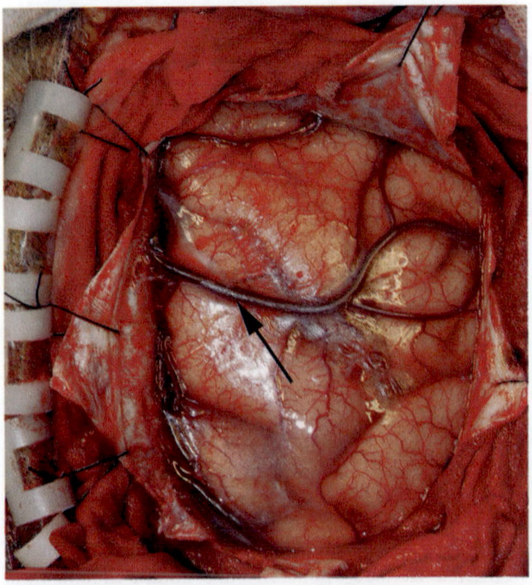

Fig. 51 Dura was opened followed by unveiling posterior frontal vein that indicated the position of nidus

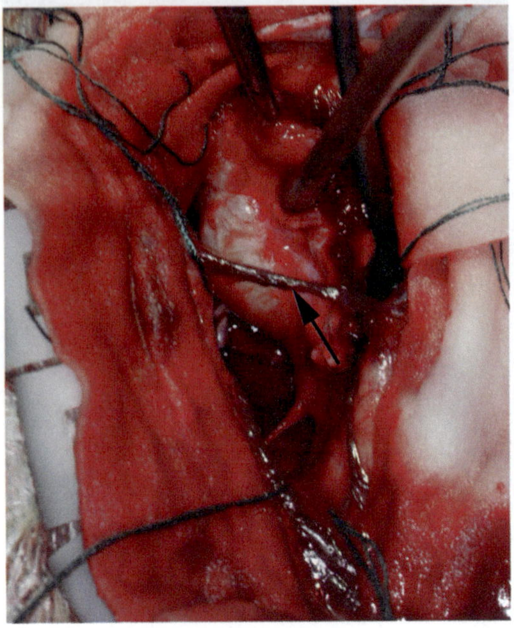

Fig. 52 The medial surface of frontal lobe was retracted laterally and mass was freed for piecemeal resection

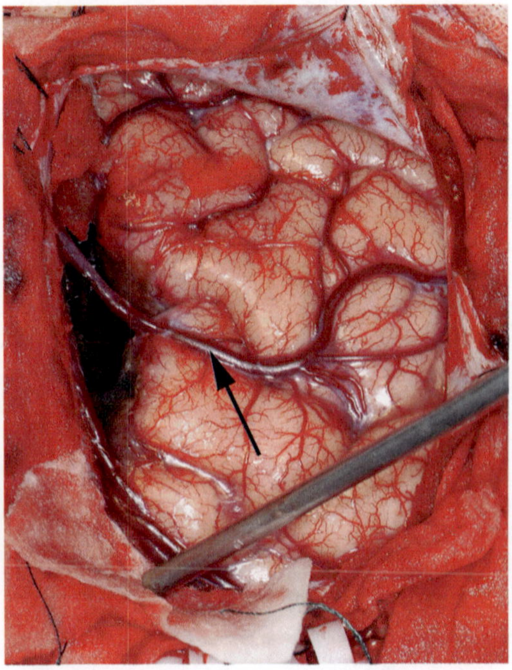

Fig. 53 Posterior frontal (arrow) and precentral vein were well preserved

Case 1.4 Right Frontal Dural Arteriovenous Fistula

Clinical presentation:

A 56-year-old lady presented with a history of dizziness and visual deficits for 20 days.

Neurological exam:

Decreased visual acuity, right quadrantanopsia

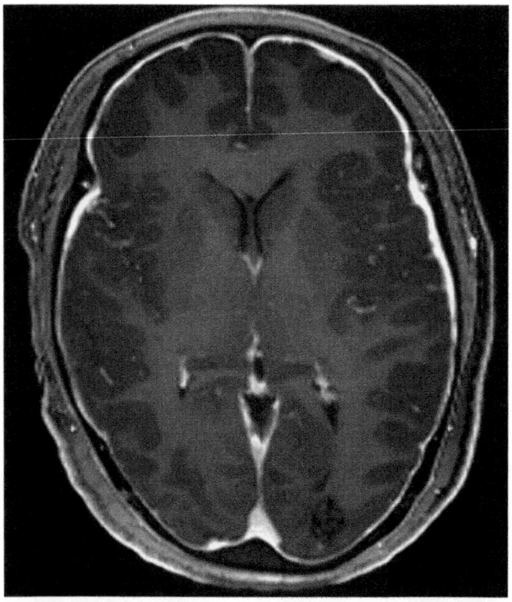

Fig. 54 Pre-op axial Gd-enhanced T1-weighted MRI

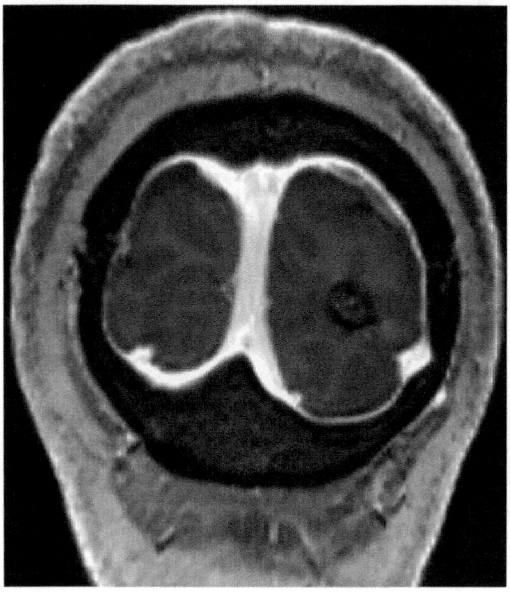

Fig. 55 Pre-op coronal Gd-enhanced T1-weighted MRI

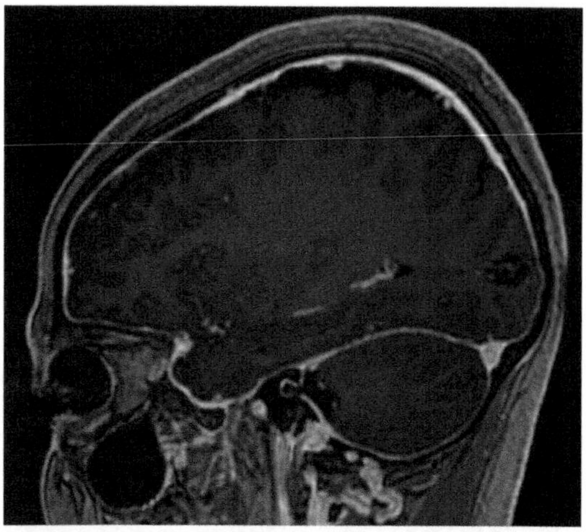

Fig. 56 Pre-op sagittal Gd-enhanced T1-weighted MRI. Pre-op Gd-enhanced T1-weighted MRI illustrated a well-circumscribed "popcorn" like nodule around right posterior calcarine sulcus without enhancement

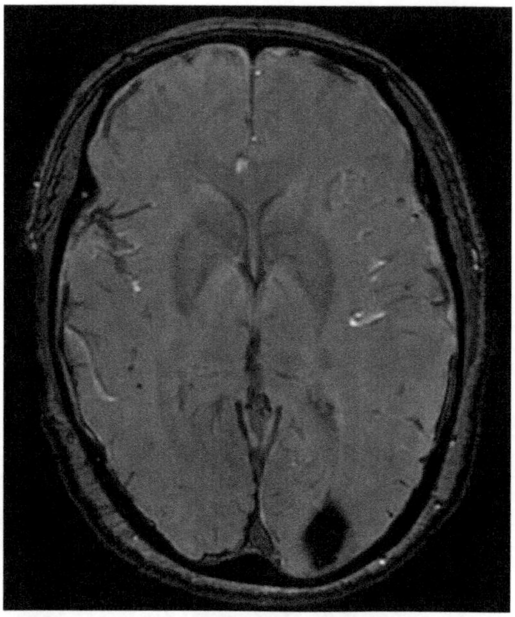

Fig. 57 Pre-op axial SWI

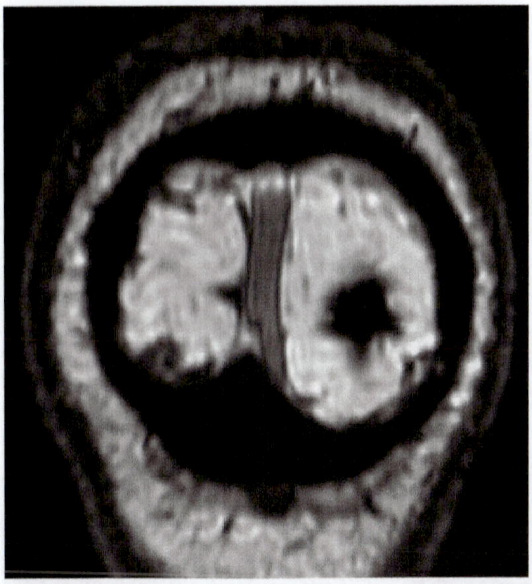

Fig. 58 Pre-op coronal SWI

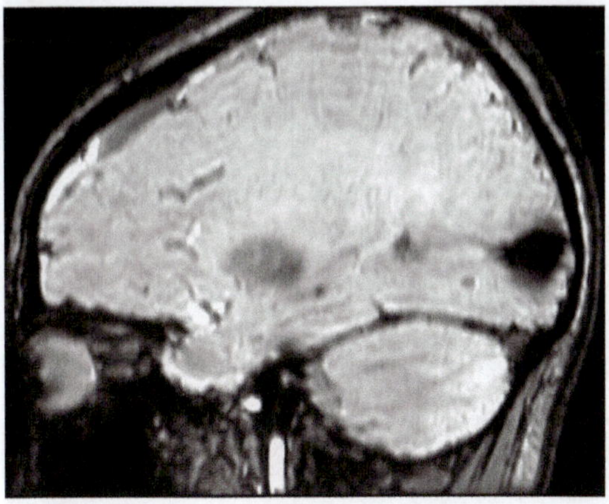

Fig. 59 Pre-op sagittal SWI. Pre-op SWI sequence confirmed hemosiderin rim that appeared as low signal intensity

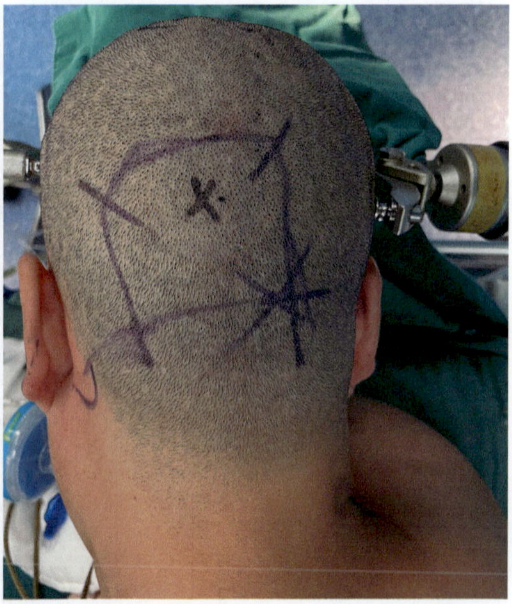

Fig. 60 She was placed in the "park bench" position and a modified parieto-occipital incision was marked. The nidus (cross) was indicated by opisthocranion, a midline skin key point

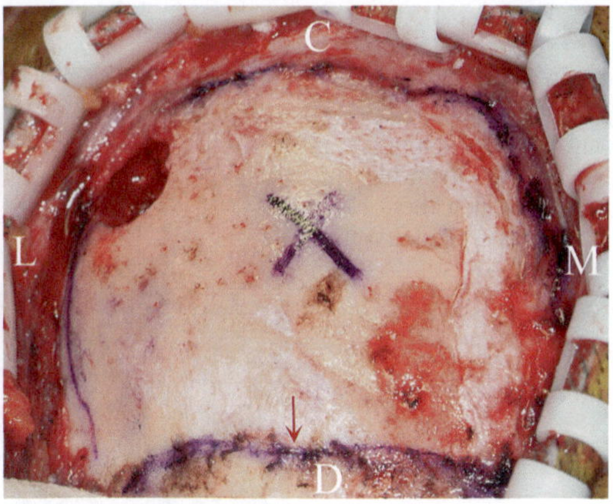

Fig. 61 Before craniotomy, superior nuchal line (arrow) was utilized to locate target lesion (cross). Only one burr hole was made at superolateral margin of bone flap

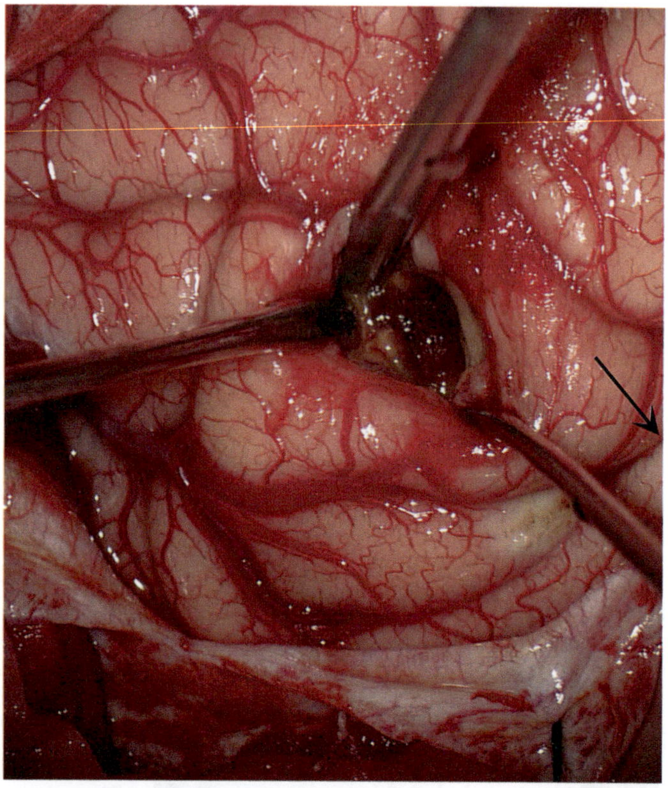

Fig. 62 The mass was approached in transcortical manner beginning at 2 cm off the midline at the level of calcarine fissure (arrow)

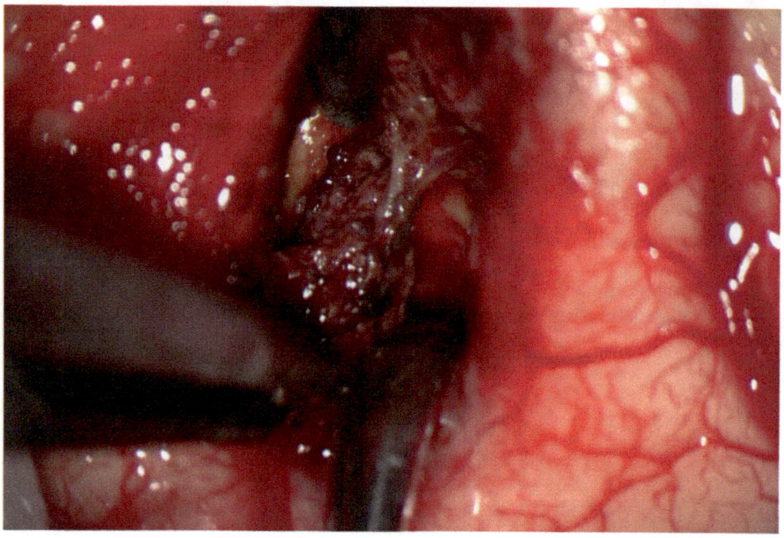

Fig. 63 The "popcorn" malformation was freed and evacuated from gliotic parenchyma by en bloc fashion after hematoma aspiration

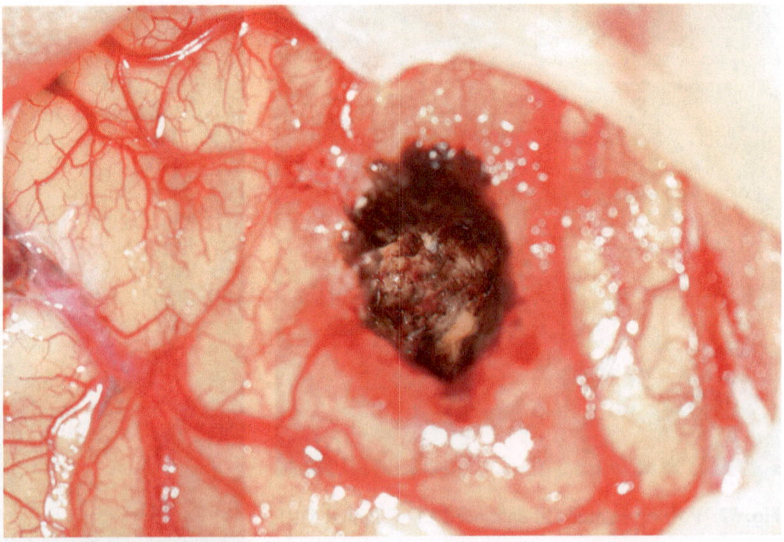

Fig. 64 Complete hemostasis was achieved by gentle coagulation of hemosiderin-laden margins

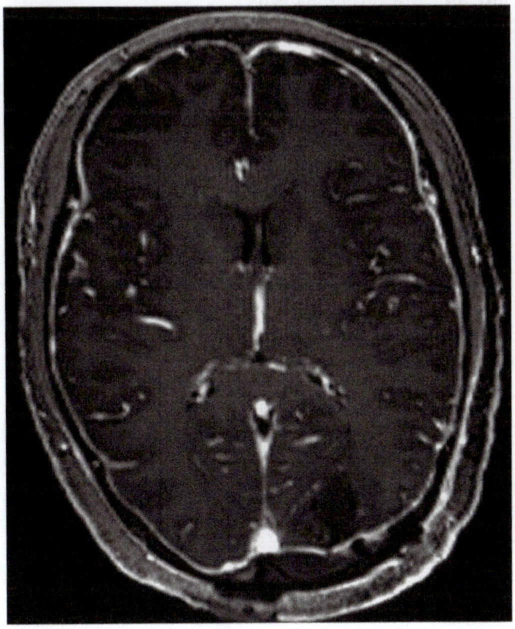

Fig. 65 Post-op axial Gd-enhanced T1-weighted MRI

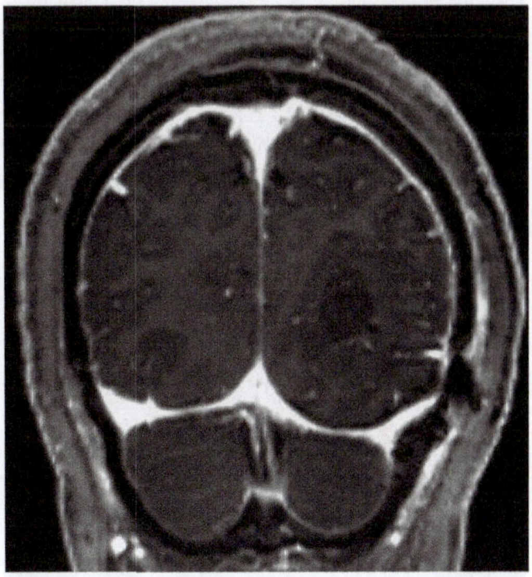

Fig. 66 Post-op coronal Gd-enhanced T1-weighted MRI

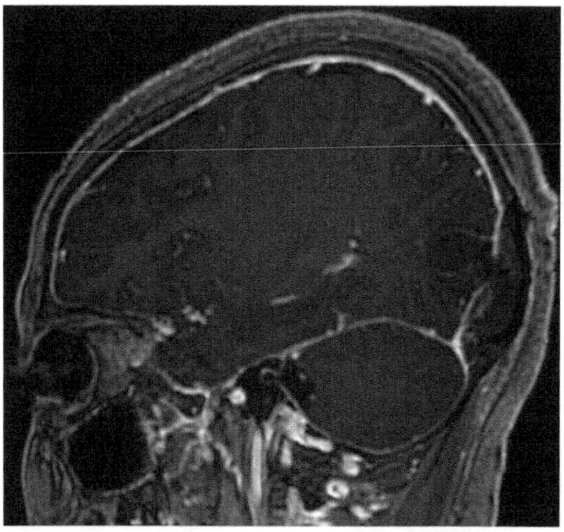

Fig. 67 Post-op sagittal Gd-enhanced T1-weighted MRI

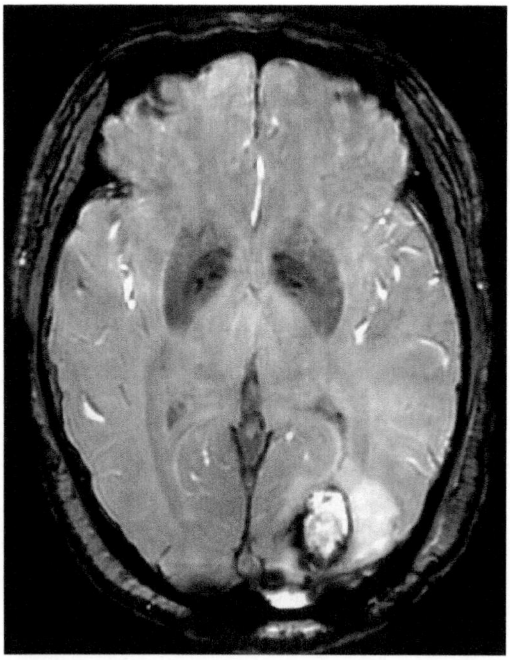

Fig. 68 Post-op axial SWI

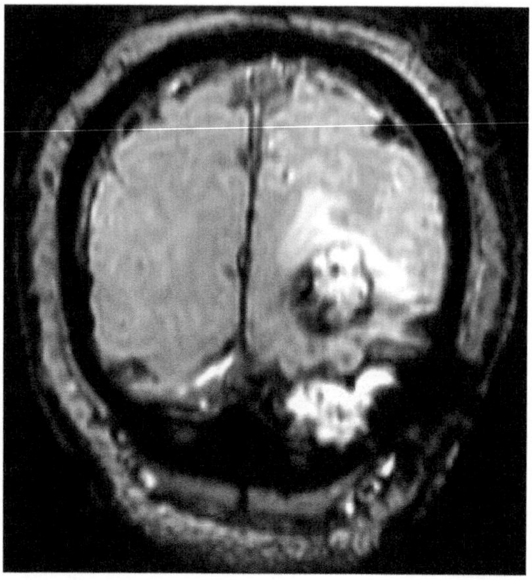

Fig. 69 Post-op coronal SWI

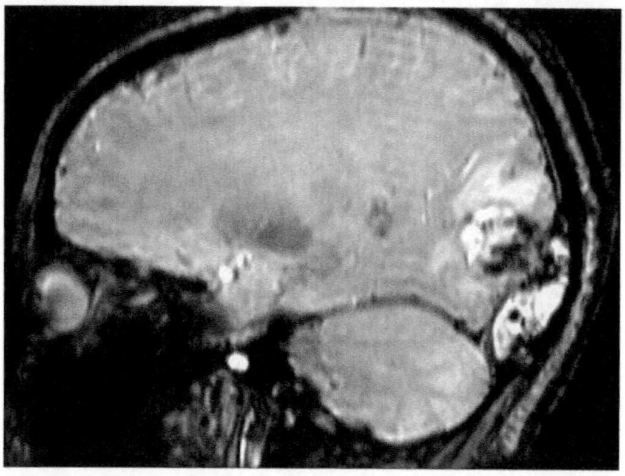

Fig. 70 Post-op sagittal SWI

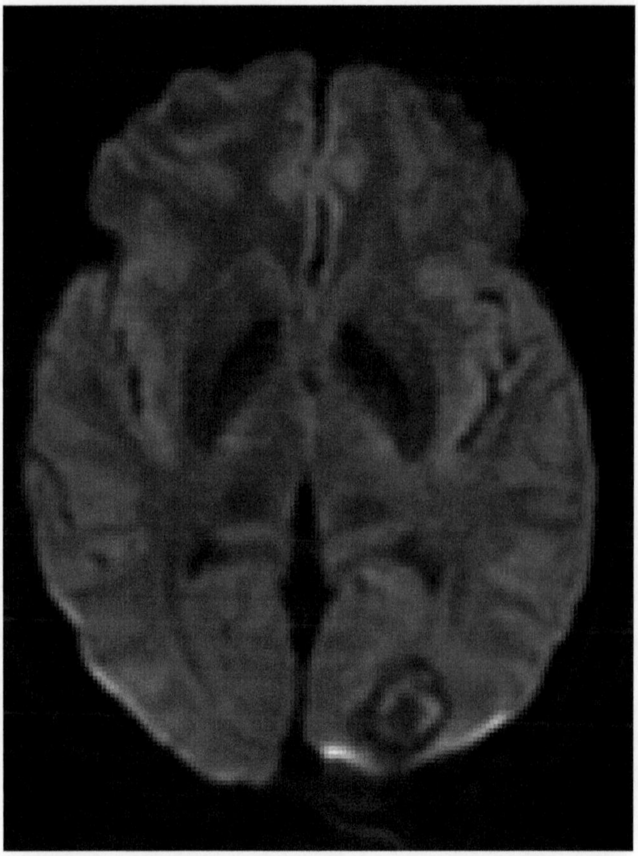

Fig. 71 Post-op DWI. Post-op MRI revealed total removal of the cavernous malformation without any new neurological deficits

Anatomy of Perisylvian Region, Basal Ganglia, Perisellar Region, Cavernous Sinus and Surgical Techniques

Xiang'En Shi, Long Wang, and Hai Qian

1 Anatomy of PeriSylvian Region (Sylvian Fissure, Perisylvian Operculum, Insular, and Relevant Vasculatures)

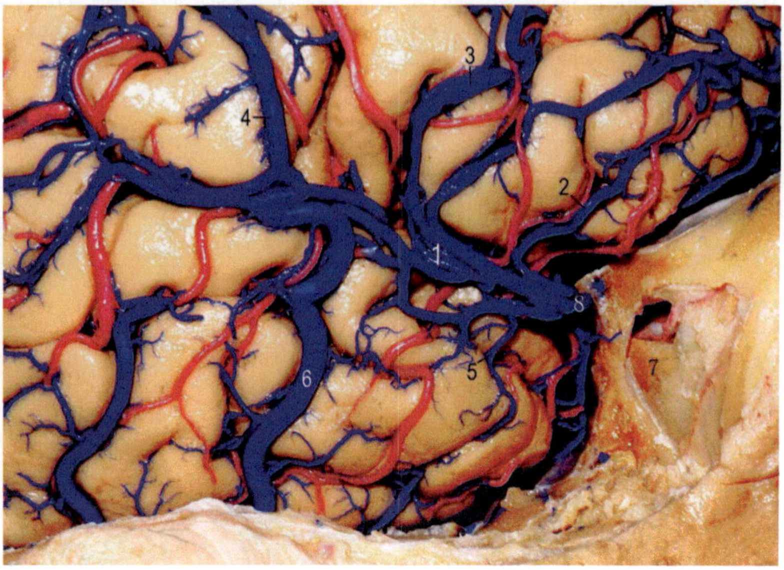

Fig. 1 Lateral view of venous structures on the surface of the right Sylvian fissure. (*1*) Superficial middle cerebral vein (Sylvian vein), (*2*) Frontopolar vein, (*3*) Middle frontal vein, (*4*) Precentral vein, (*5*) Temporal sylvian vein, (*6*) Middle temporal vein, (*7*) Lateral rectus muscle, (*8*) Temporopolar bridging vein

X. Shi (✉) · L. Wang · H. Qian
Neurological Surgery, SanBo Brain Hospital, Capital Medical University, Beijing, China
e-mail: shixen@ccmu.edu.cn;
wanglong@ccmu.edu.cn; qianh@ccmu.edu.cn

X. Shi et al. (eds.), *Microneurosurgical Anatomy and Surgical Technique*,
https://doi.org/10.1007/978-981-19-8273-6_2

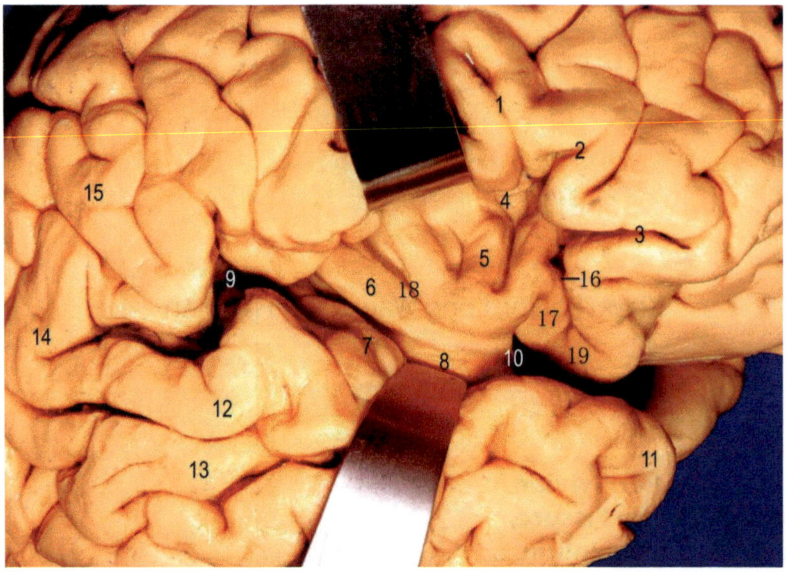

Fig. 2 Lateral view of right perisylvian operculum and insula. (*1*) Ascending ramus, (*2*) Horizontal ramus, (*3*) Frontoorbital ramus, (*4*) Superior limiting sulcus, (*5*) Middle short gyrus of insula, (*6*) Anterior long gyrus of insula, (*7*) Transverse temporal gyrus, (*8*) Inferior limiting sulcus, (*9*) Posterior ramus of lateral sulcus, (*10*) Limen insula, (*11*) Temporal pole, (*12*) Superior temporal gyrus, (*13*) Middle temporal gyrus, (*14*) Angular gyrus, (*15*) Supramarginal gyrus, (*16*) Anterior limen sulcus, (*17*) Accessary gyrus, (*18*) Central sulcus of insula, (*19*) Transverse insular gyrus

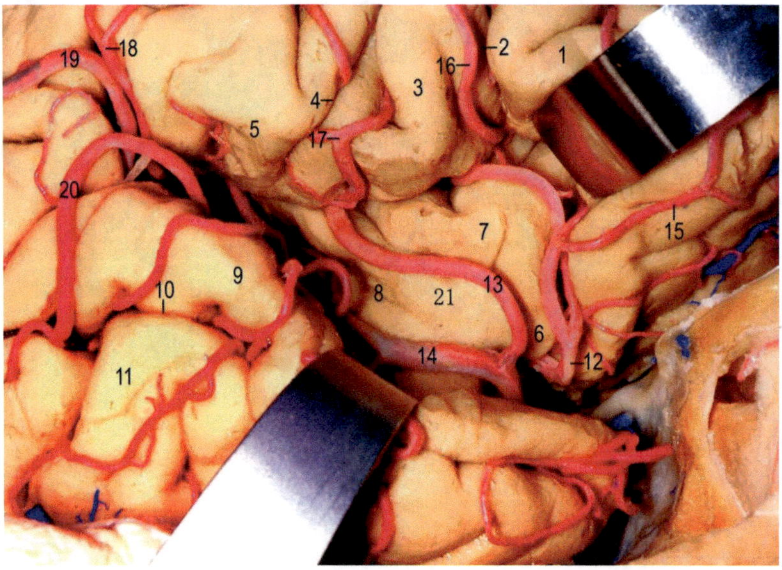

Fig. 3 Lateral view arterial structures covering right perisylvian cortex and insula. (*1*) Pars Triangularis, (*2*) Ascending ramus, (*3*) Pars opercularis, (*4*) Precentral sulcus, (*5*) Subcentral gyrus, (*6*) Insular pole, (*7*) Middle short gyrus of insula, (*8*) Central insular sulcus, (*9*) Superior temporal gyrus, (*10*) Superior temporal sulcus, (*11*) Middle temporal gyrus, (*12*) Early frontal branch, (*13*) Frontal trunk, (*14*) Temporal trunk, (*15*) Lateral Orbitofrontal artery, (*16*) Prefrontal artery, (*17*) Precentral artery, (*18*) Anterior and posterior parietal artery, (*19*) Angular artery, (*20*) Temporooccipital artery, (*21*) Posterior short gyrus of insula

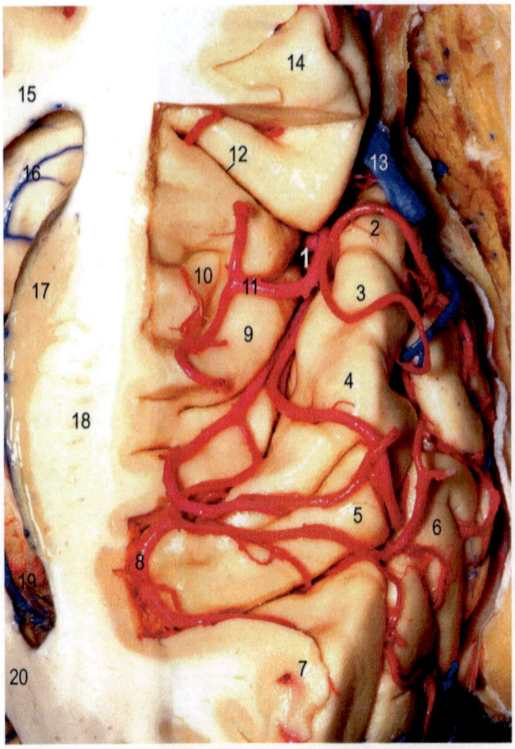

Fig. 4 Superior view of right perisylvian operculum and insula. (*1*) Sylvian fissure, (*2*) Temporal polar, (*3*) Planum polar, (*4*) Anterior temporal transverse gyrus, (*5*) Middle temporal transverse gyrus, (*6*) Superior temporal gyrus, (*7*) Parietal operculum, (*8*) Angular artery, (*9*) Middle short gyrus of insula, (*10*) Short insular sulcus, (*11*) Prefrontal artery, (*12*) Anterior limiting sulcus of insula, (*13*) Temporopolar bridging vein, (*14*) Frontal operculum, (*15*) Genu of corpus callosum, (*16*) Frontal horn, (*17*) Head of caudate nucleus, (*18*) Corona radiata, (*19*) Choroid plexus, (*20*) Splenium of corpus callosum

2 Anatomy of Basal Ganglia

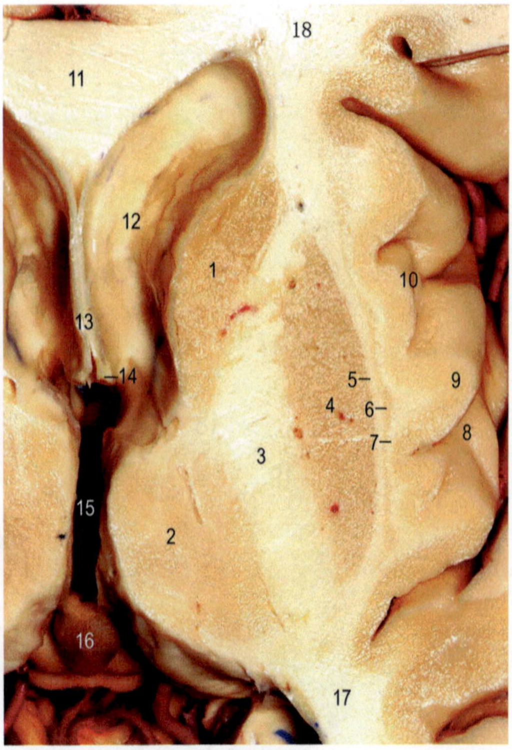

Fig. 5 Superior view of right basal ganglia. (*1*) Head of caudate nucleus, (*2*) Thalamus, (*3*) Internal capsule, (*4*) Putamen, (*5*) External capsule, (*6*) Claustrum, (*7*) Extreme capsule, (*8*) Anterior long gyrus of insula, (*9*) Insular cortex, (*10*) Middle short gyrus of insula, (*11*) Genu of corpus callosum, (*12*) Frontal horn of lateral ventricle, (*13*) Septum pellucidum, (*14*) Column of fornix, (*15*) Third ventricle, (*16*) Pineal gland, (*17*) Posterior isthmus, (*18*) Anterior isthmus

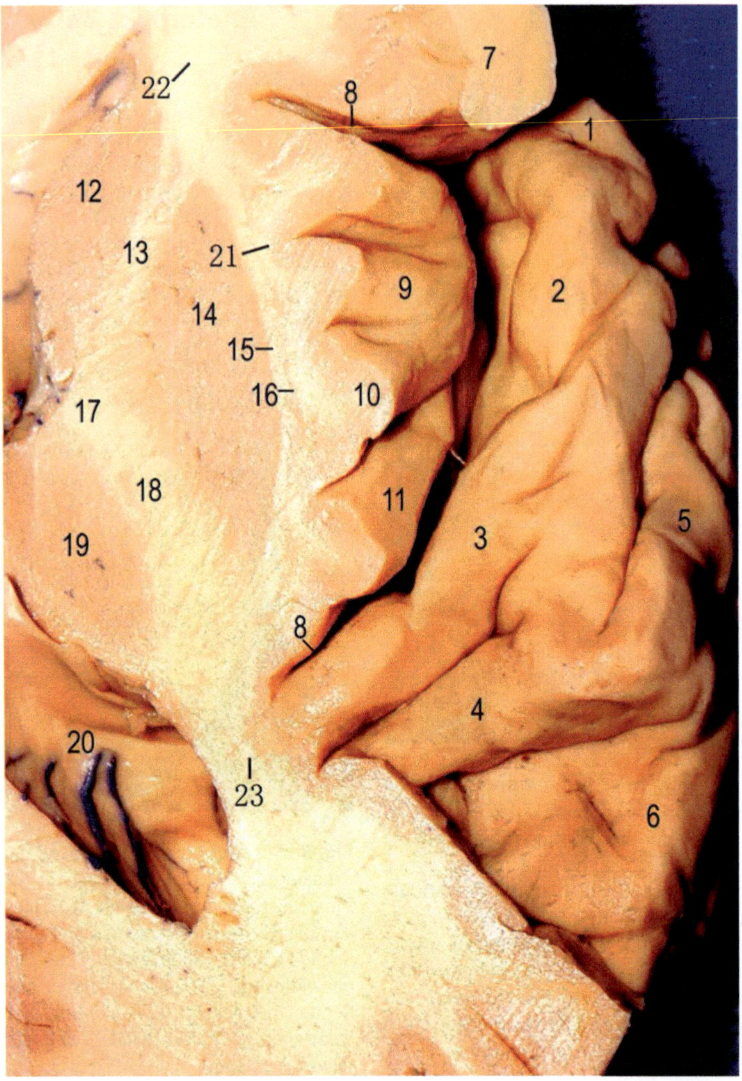

Fig. 6 Axial view of right basal ganglia, temporal operculum, and insular cortex. (*1*) Temporal pole, (*2*) Planum polare, (*3*) Anterior temporal transverse gyrus, (*4*) Middle temporal transverse gyrus, (*5*) Superior temporal gyrus, (*6*) Posteriro temporal transverse gyrus, (*7*) Frontal lobe, (*8*) Anterior insular limiting sulcus, (*9*) Middle short gyrus of insula, (*10*) Insular cortex, (*11*) Anterior long gyrus of insula, (*12*) Head of caudate nucleus, (*13*) Anterior limb of internal capsule, (*14*) Putamen, (*15*) External capsule, (*16*) Claustrum, (*17*) Genu of internal capsule, (*18*) Posterior limb of internal capsule, (*19*) Thalamus, (*20*) Trigone of lateral ventricle, (*21*) Extreme capsule, (*22*) Anterior isthmus, (*23*) Posterior isthmus

3 Anatomy of Perisellar (Sellar, Suprasellar, and Parasellar) Region

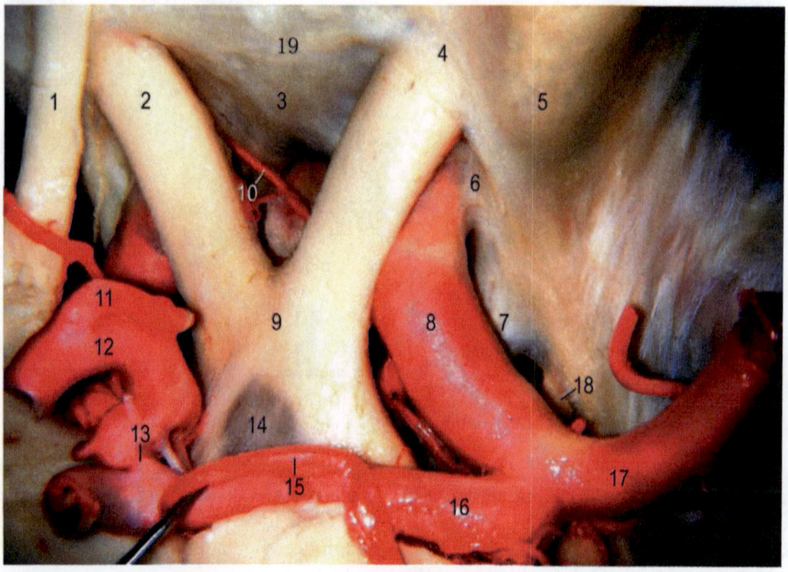

Fig. 7 Overview of perisellar region. (*1*) Olfactory tract, (*2*) Left optic nerve, (*3*) Tuberculum sellae, (*4*) Falciform ligament, (*5*) Right anterior clinoid process, (*6*) Right upper (distal) dural ring, (*7*) Posterior clinoid process, (*8*) Internal carotid artery, (*9*) Optic chiasm, (*10*) Superior hypophyseal artery, (*11*) Left temporopolar artery, (*12*) Left middle cerebral artery, (*13*) Anterior communicating artery, (*14*) Lamina terminalis, (*15*) Right recurrent artery of Heubner, (*16*) Right anterior cerebral artery, (*17*) Right middle cerebral artery, (*18*) Oculomotor nerve, (*19*) Chiasmatic sulcus

Fig. 8 Opening of lamina terminalis. (*1*) Optic chiasm, (*2*) Right internal carotid artery, (*3*) Left recurrent artery of Heubner, (*4*) Third ventricle, (*5*) Fenestration of anterior communicating artery, (*6*) Right anterior cerebral artery, (*7*) Right recurrent artery of Heubner

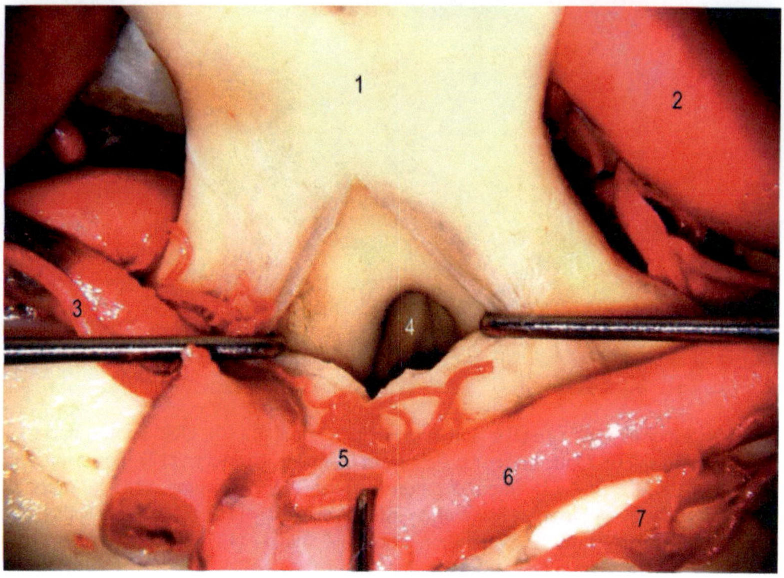

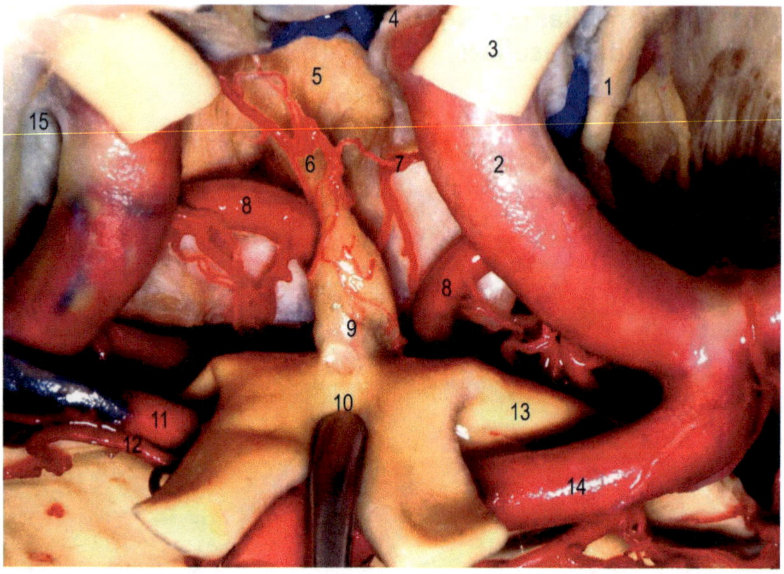

Fig. 9 Anterior view of perisellar region. (*1*) Oculomotor nerve, (*2*) Internal carotid artery, (*3*) Optic nerve, (*4*) Right upper (distal) dural ring, (*5*) Pituitary gland, (*6*) Pituitary stalk, (*7*) Superior hypophysial artery, (*8*) Posterior communicating artery, (*9*) Infundibulum, (*10*) Optic chiasm, (*11*) Precommunicating segment of left anterior cerebral artery, (*12*) Recurrent artery of Heubner, (*13*) Optic tract, (*14*) Precommunicating segment of right anterior cerebral artery, (*15*) Left upper (distal) dural ring

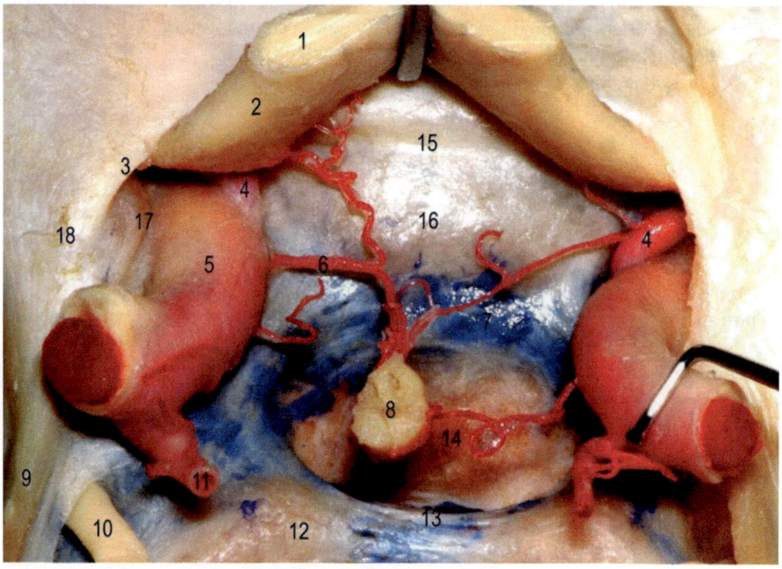

Fig. 10 Superior view of sellar and suprasellar region. (1) Optic chiasm, (2) Optic nerve, (3) Falciform ligament, (4) Ophthalmic artery, (5) Internal carotid artery, (6) Superior hypophysial artery, (7) Anterior intercavernous sinus, (8) Pituitary stalk, (9) Anterior petroclinoid fold, (10) Oculomotor nerve, (11) Posterior communicating artery, (12) Posterior clinoid process, (13) Sella diaphragma, (14) Pituitary gland, (15) Chiasmatic sulcus, (16) Tuberculum sellae, (17) Left upper (distal) dural ring, (18) Anterior clinoid process

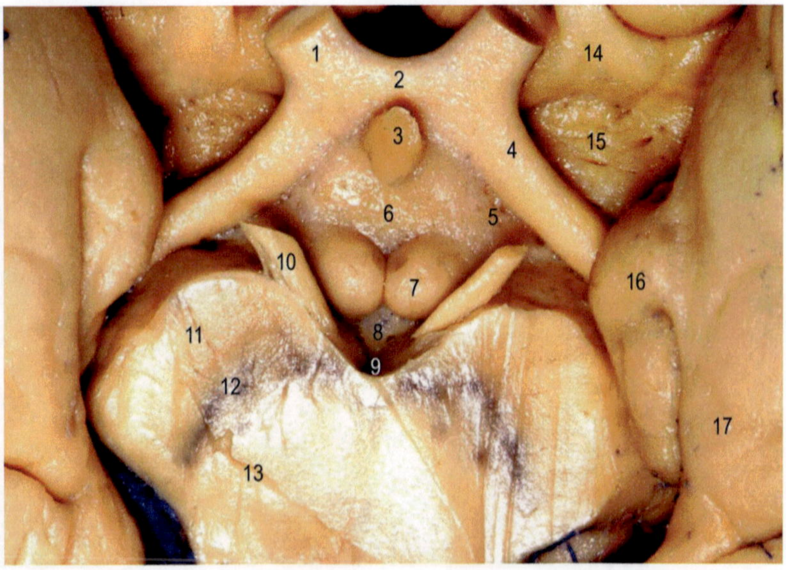

Fig. 11 Inferior view of suprasellar region. (*1*) Optic nerve, (*2*) Optic chiasm, (*3*) Infundibulum, (*4*) Optic tract, (*5*) Middle perforated substance, (*6*) Tuber cinereum, (*7*) Mamillary body, (*8*) Posterior perforated substance, (*9*) Interpeduncular fossa, (*10*) Oculomotor nerve, (*11*) Cerebral peduncle, (*12*) Substantia nigra, (*13*) Red nucleus, (*14*) Olfactory trigone, (*15*) Anterior perforated substance, (*16*) Uncus, (*17*) Parahippocampal gyrus

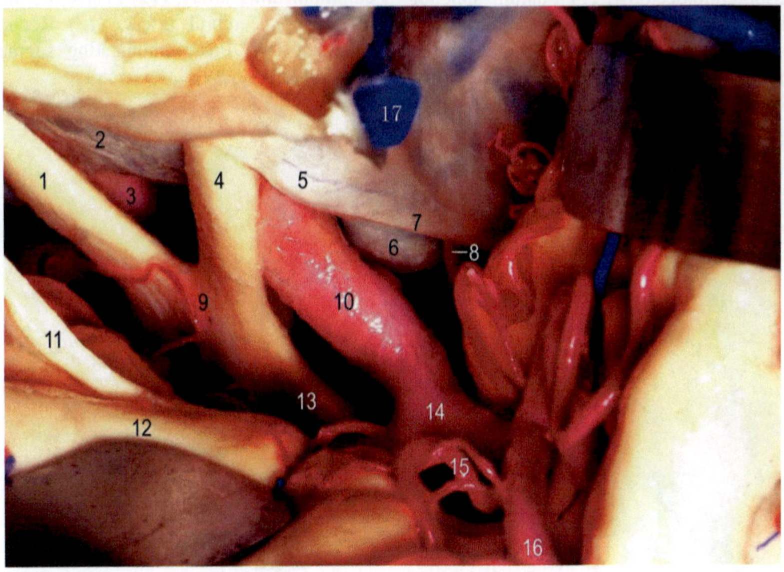

Fig. 12 Overview of right parasellar region. (*1*) Left optic nerve, (*2*) Tuberculum sellae, (*3*) Left internal carotid artery, (*4*) Right optic nerve, (*5*) Anterior clinoid process, (*6*) Posterior clinoid process, (*7*) Anterior petroclinoid fold, (*8*) Oculomotor nerve, (*9*) Optic chiasm, (*10*) Right internal carotid artery, (*11*) Olfactory tract, (*12*) Posterior orbital gyrus, (*13*) Optic tract, (*14*) Bifurcation of internal carotid artery, (*15*) Early branch of middle cerebral artery, (*16*) Middle cerebral artery, (*17*) Sphenoparietal sinus

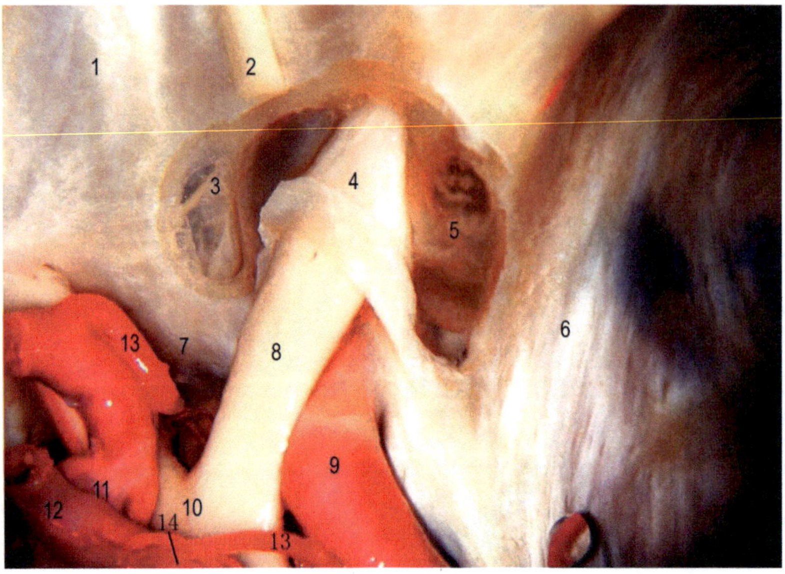

Fig. 13 Superior view of right parasellar region with intradural anterior clinoidectomy. (*1*) Planum sphenoidale, (*2*) Olfactory tract, (*3*) Mucosa of sphenoid sinus, (*4*) Falciform ligament, (*5*) Optic strut, (*6*) Lateral wall of cavernous sinus, (*7*) Tuberculum sellae, (*8*) Optic nerve, (*9*) Internal carotid artery, (*10*) Optic chiasm, (*11*) Anterior communicating artery, (*12*) Postcommunicating segment of anterior cerebral artery, (*13*) Recurrent artery of Heubner, (*14*) Precommunicating segment of anterior cerebral artery

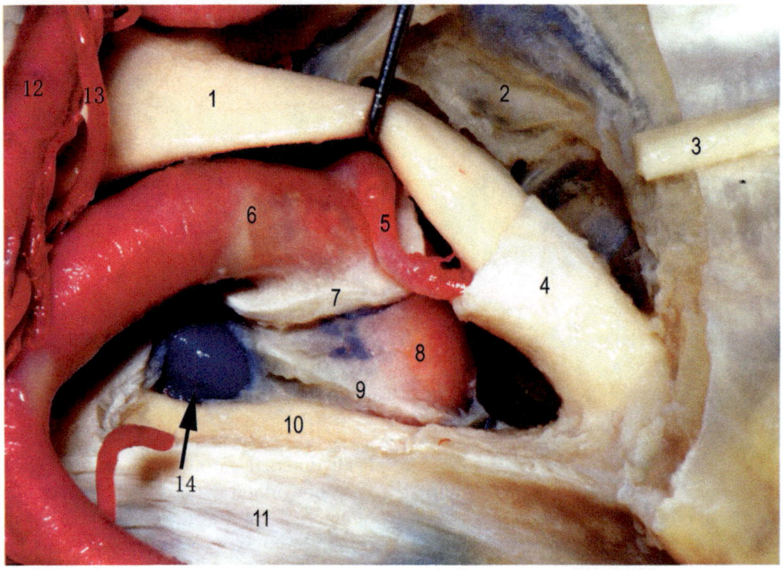

Fig. 14 Lateral view of right parasellar region. (*1*) Optic nerve, (*2*) Mucosa of sphenoid sinus, (*3*) Olfactory tract, (*4*) Falciform ligament, (*5*) Ophthalmic artery, (*6*) Internal carotid artery, (*7*) Upper (Distal) dural ring, (*8*) Clinoid segment of internal carotid artery, (*9*) Inferior (Proximal) dural ring, (*10*) Oculomotor nerve, (*11*) Lateral wall of cavernous sinus, (*12*) Precommunicating segment of anterior cerebral artery, (*13*) Recurrent artery of Heubner, (*14*) Cavernous sinus

4 Anatomy of Cavernous Sinus

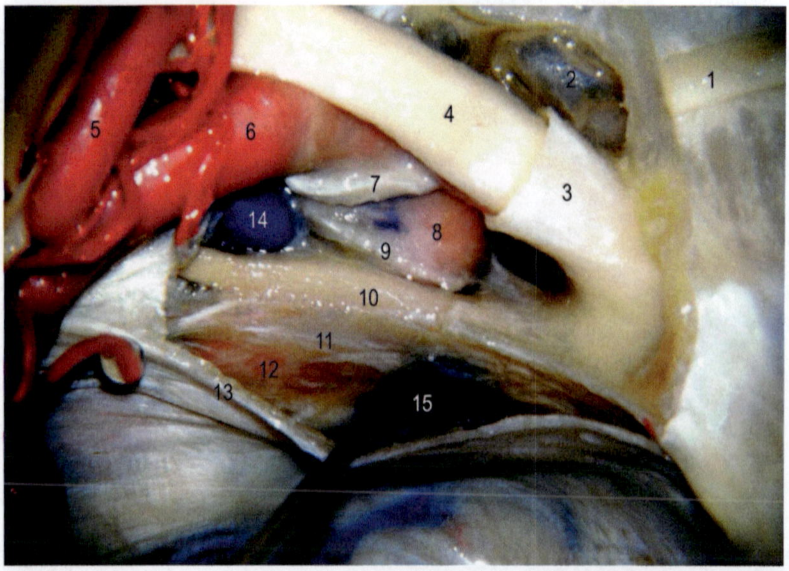

Fig. 15 Lateral view of lateral wall of right cavernous sinus. (*1*) Olfactory tract, (*2*) Mucosa of sphenoid sinus, (*3*) Falciform ligament, (*4*) Optic nerve, (*5*) Anterior cerebral artery, (*6*) Internal carotid artery, (*7*) Upper (Distal) dural ring, (*8*) Clinoid segment of internal carotid artery, (*9*) Inferior (Proximal) dural ring, (*10*) Oculomotor nerve, (*11*) Periosteal (internal) layer of lateral wall of cavernous sinus, (*12*) Cavernous segment of internal carotid artery, (*13*) Meningeal (external) layer of lateral wall of cavernous sinus, (*14*) Cavernous sinus, (*15*) Interspace of lateral wall of cavernous sinus

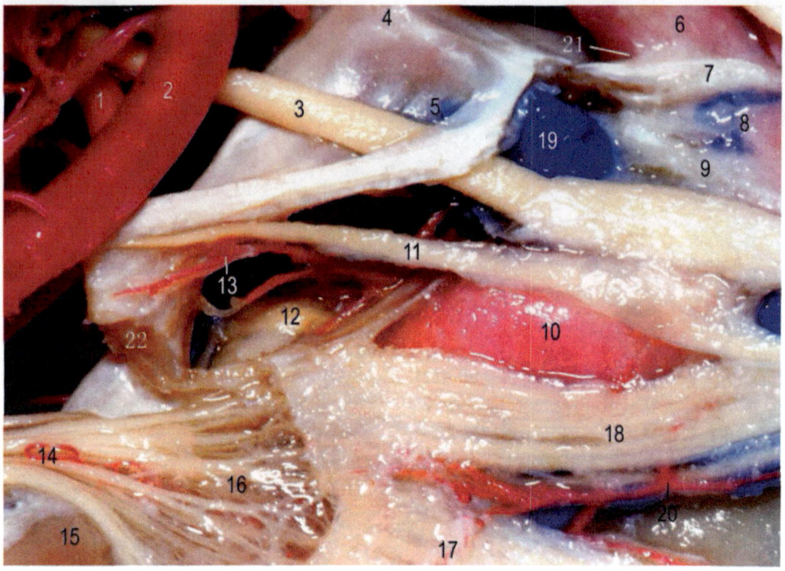

Fig. 16 Lateral overview of cavernous sinus. (*1*) Superior cerebellar artery, (*2*) Posterior cerebral artery, (*3*) Intracranial segment of oculomotor nerve, (*4*) Posterior clinoid process, (*5*) Porus of oculomotor nerve, (*6*) Ophthalmic segment of internal carotid artery, (*7*) Upper (Distal) dural ring, (*8*) Clinoid segment of internal carotid artery, (*9*) Inferior (Proximal) dural ring, (*10*) Cavernous segment of internal carotid artery, (*11*) Trochlear nerve, (*12*) Abducens nerve, (*13*) Marginal tentorial artery (artery of Bernasconi and Cassinari), (*14*) Trigeminal trunk, (*15*) Petrous apex, (*16*) Meckel's cave, (*17*) Maxillary nerve, (*18*) Ophthalmic nerve, (*19*) Cavernous sinus, (*20*) Inferolateral trunk, (*21*) Carotid cave, (*22*) Tentorial incisura

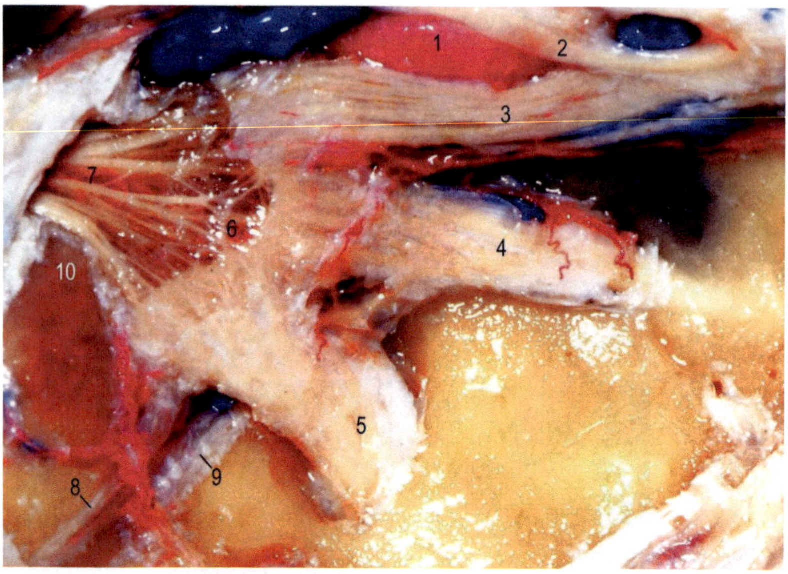

Fig. 17 Lateral view of intracavernous structures. (*1*) Cavernous segment of internal carotid artery, (*2*) Trochlear nerve, (*3*) Ophthalmic nerve, (*4*) Maxillary nerve, (*5*) Mandibular nerve, (*6*) Meckel's cave, (*7*) Root of trigeminal nerve, (*8*) Greater superficial petrosal nerve, (*9*) Lesser petrosal nerve, (*10*) Petrous apex

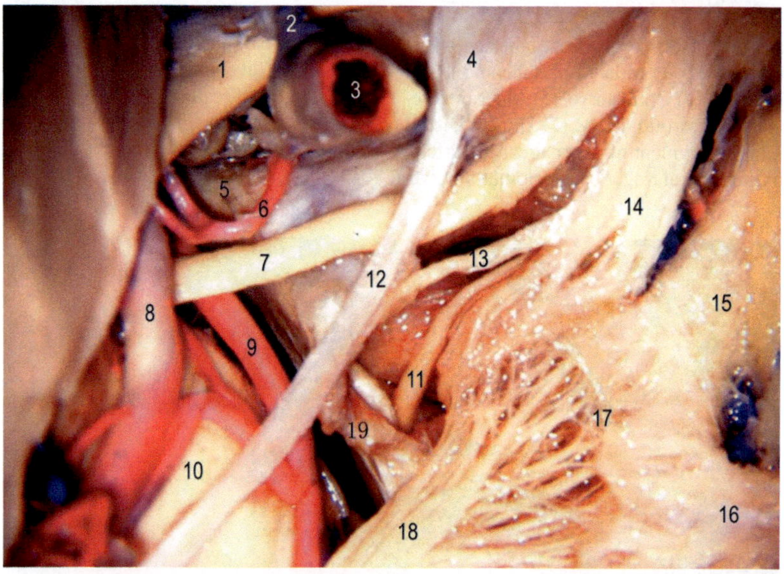

Fig. 18 Superior view of cavernous sinus. (*1*) Optic tract, (*2*) Anterior cerebral artery, (*3*) Middle cerebral artery, (*4*) Anterior clinoid process, (*5*) Dorsum sellae, (*6*) Posterior communicating artery, (*7*) Oculomotor nerve, (*8*) Posterior cerebral artery, (*9*) Basilar artery, (*10*) Pons, (*11*) Abducens nerve, (*12*) Anterior petroclinoid fold, (*13*) Trochlear nerve, (*14*) Ophthalmic nerve, (*15*) Maxillary nerve, (*16*) Mandibular nerve, (*17*) Trigeminal ganglion, (*18*) Posterior root of trigeminal nerve, (*19*) Posterior petroclinoid fold

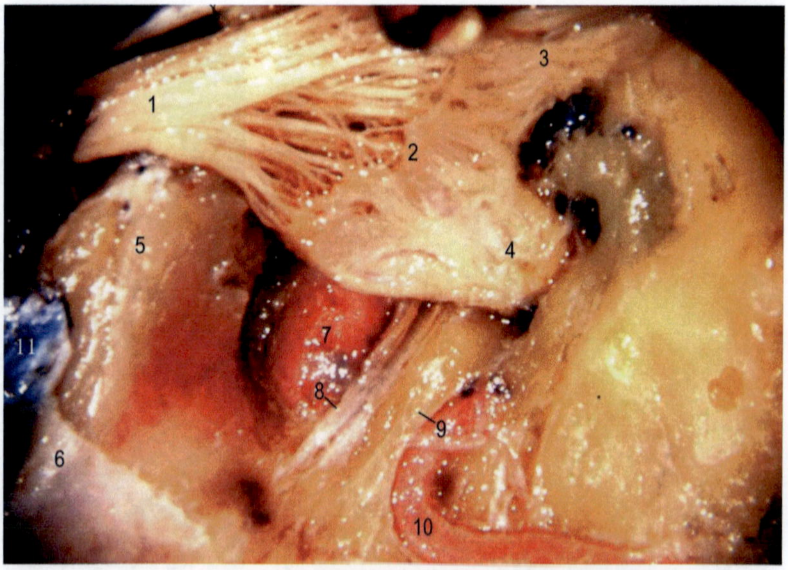

Fig. 19 Posterior view of cavernous sinus. (*1*) Posterior root of trigeminal nerve, (*2*) Trigeminal ganglion, (*3*) Maxillary nerve, (*4*) Mandibular nerve, (*5*) Petrous apex, (*6*) Petrous ridge, (*7*) Petrous segment of internal carotid artery, (*8*) Greater superficial petrosal nerve, (*9*) Lesser superficial petrosal nerve, (*10*) Middle meningeal artery, (*11*) Superior petrosal sinus

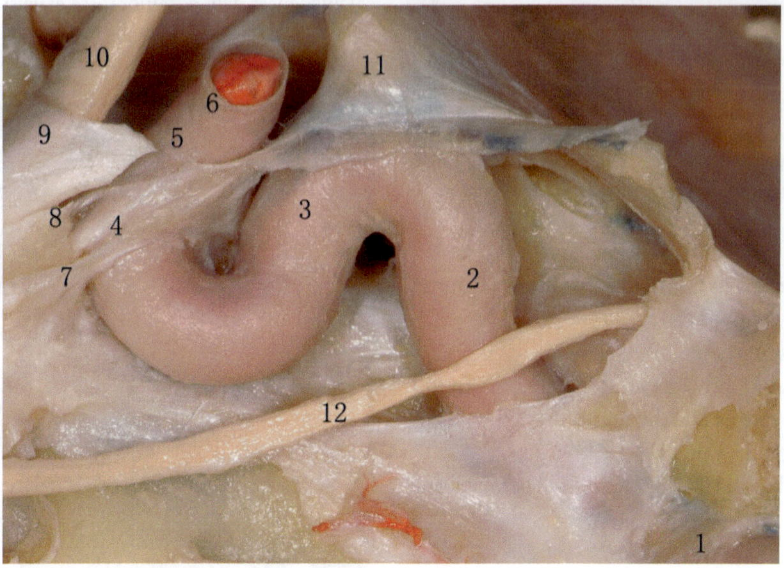

Fig. 20 Lateral view of cavernous segment of internal carotid artery. (*1*) Petrous segment of internal carotid artery, (*2*) Posterior vertical segment of cavernous internal carotid artery, (*3*) Clinoid segment of internal carotid artery, (*4*) Inferior clinoid segment of internal carotid artery, (*5*) Ophthalmic segment of internal carotid artery, (*6*) Posterior communicating segment of internal carotid artery, (*7*) Inferior (Proximal) dural ring, (*8*) Upper (Distal) dural ring, (*9*) Falciform ligament Optic nerve, (*10*) Posterior clinoid process, (*11*) Abducens nerve

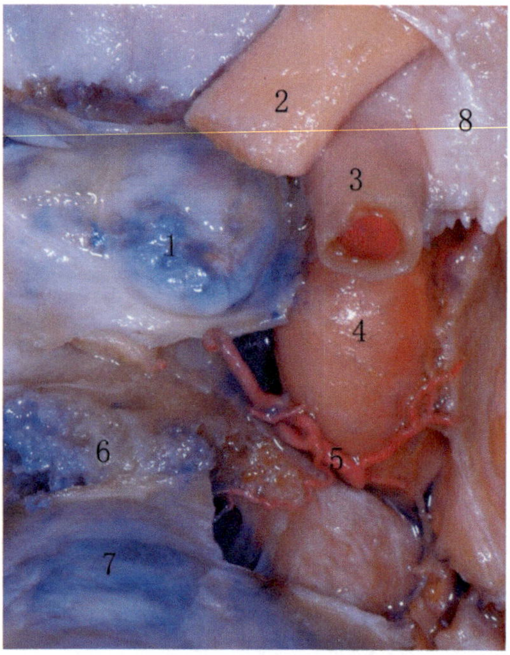

Fig. 21 Posterior view of cavernous segment of internal carotid artery. (*1*) Posterior intercavernous sinus, (*2*) Optic nerve, (*3*) Supraclinoid segment of internal carotid artery, (*4*) Posterior bend of cavernous internal carotid artery, (*5*) Meningohypophyseal trunk, (*6*) Dorsum sellae, (*7*) Basilar venous plexus, (*8*) Anterior clinoid process

Case 2.1 Pediatric Giant Craniopharyngioma (Subfrontal Approach)

Clinical presentation:

A 6-year-old boy presented with a 3-month history of severe headache and blurry vision.

Neurological exam:

Decreased visual acuity, bitemporal hemianopsia

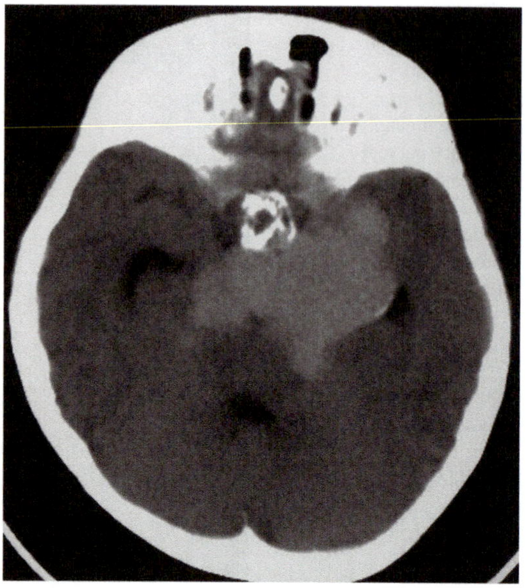

Fig. 22 Pre-op CT scan demonstrated a giant perisellar calcified mass with posterior fossa extension. The temporal horn has been deformed by this lesion

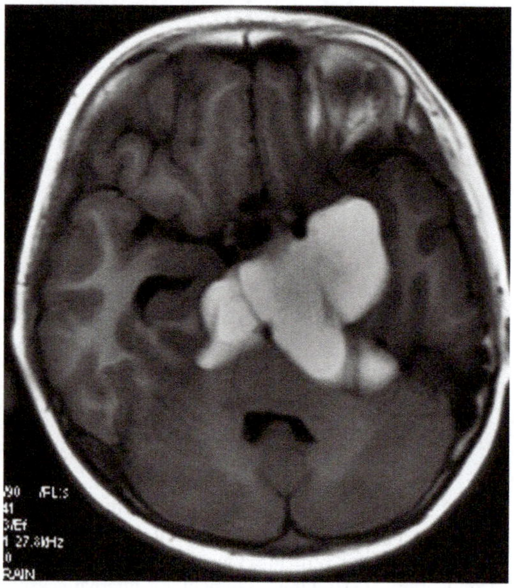

Fig. 23 Pre-op axial T1-weighted MRI

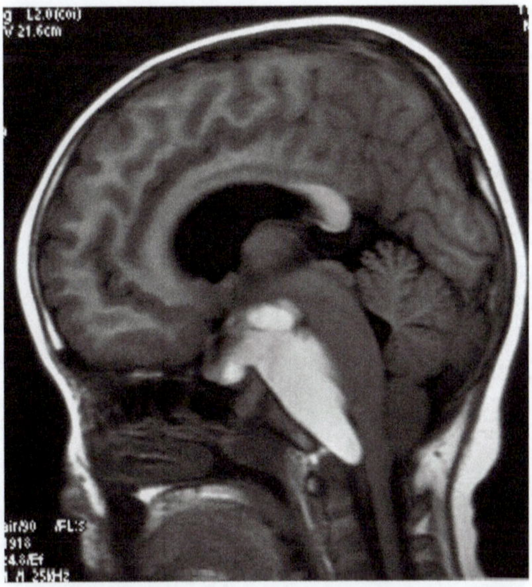

Fig. 24 Pre-op sagittal T1-weighted MRI. Pre-op MR images revealed a giant cystic suprasellar lesion with extension into parasellar region, clivus, temporal lobe, and posterior fossa

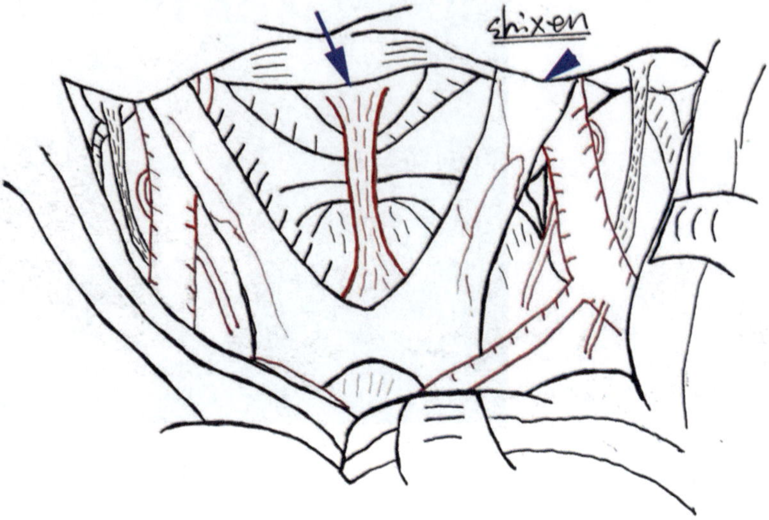

Fig. 25 The medical illustration showing exposed structures such as pituitary stalk (arrow) and right optic nerve (arrowhead) via subfrontal approach

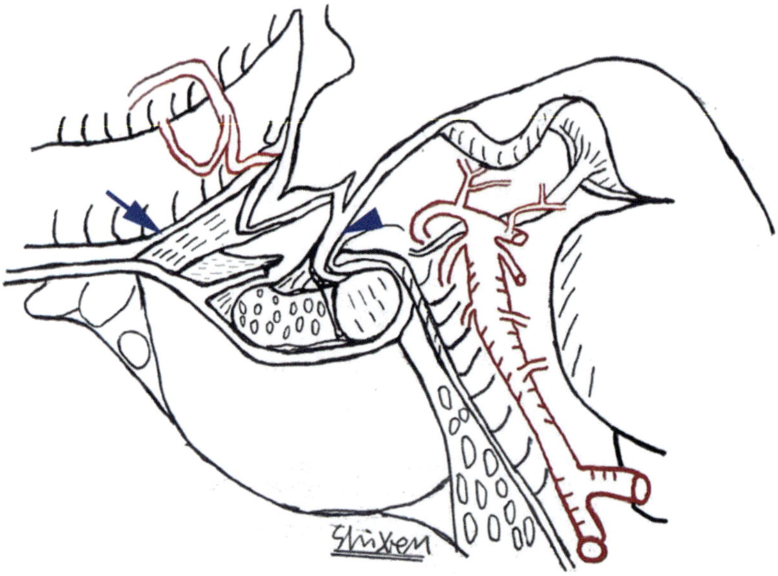

Fig. 26 The illustration from sagittal view described position of pituitary stalk (arrowhead) and optic nerve (arrow)

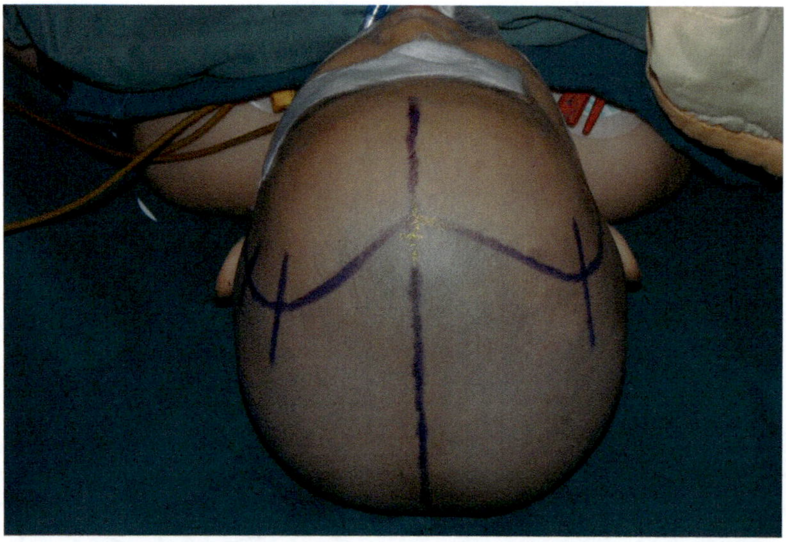

Fig. 27 The child was placed in supine position. Bifrontal skin incision was marked

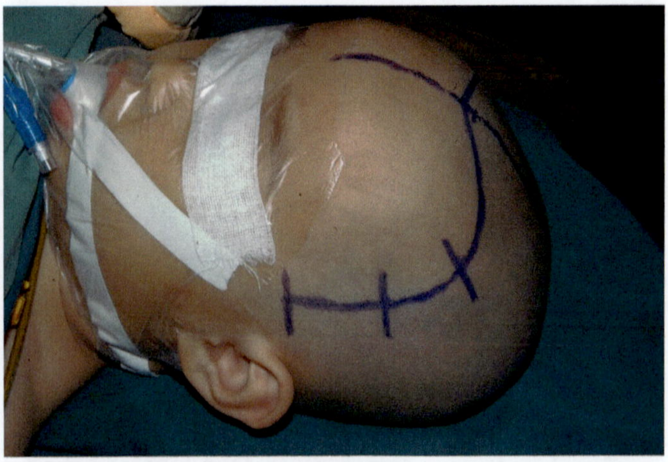

Fig. 28 Skin incision was extended into the level of geison

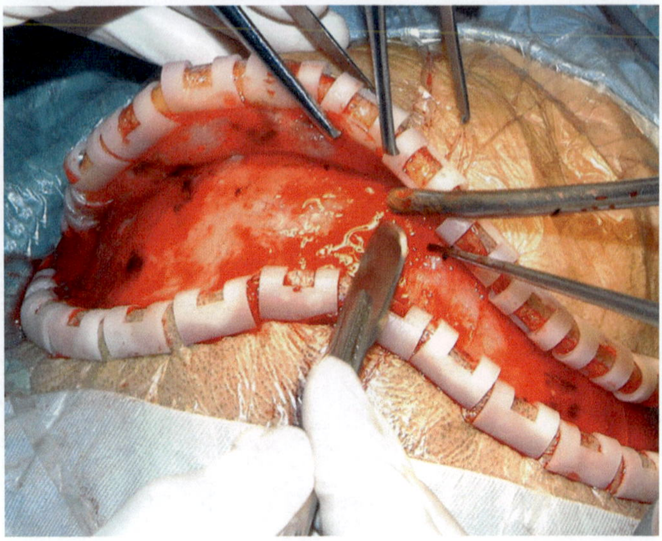

Fig. 29 Flap was created by subgaleal fashion

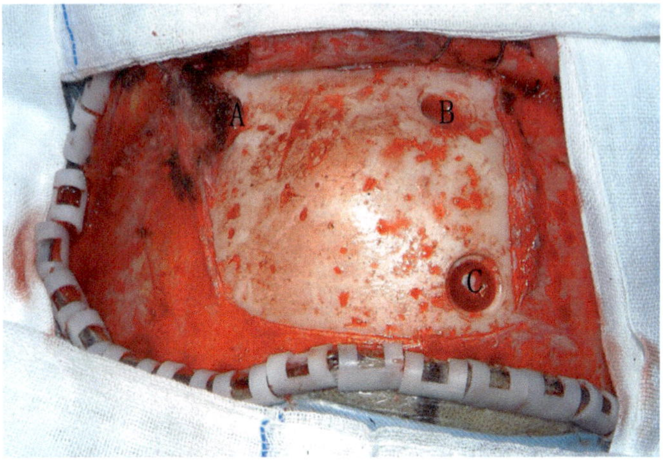

Fig. 30 Three burr holes were made at left frontal region. One was placed posterior to frontozygomatic suture (*A*) while other two were drilled over the sagittal sinus (*B* & *C*)

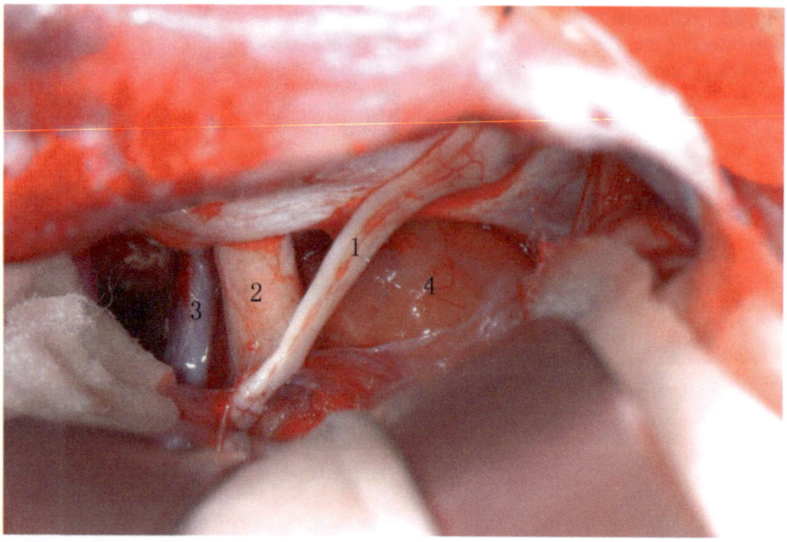

Fig. 31 Tumor was easily exposed by superior retraction of basal surface of frontal lobe. (*1*) ipsilateral olfactory tract, (*2*) optic nerve, (*3*) supraclinoidal segment of internal carotid artery, (*4*) tumor (via prechiasmatic space)

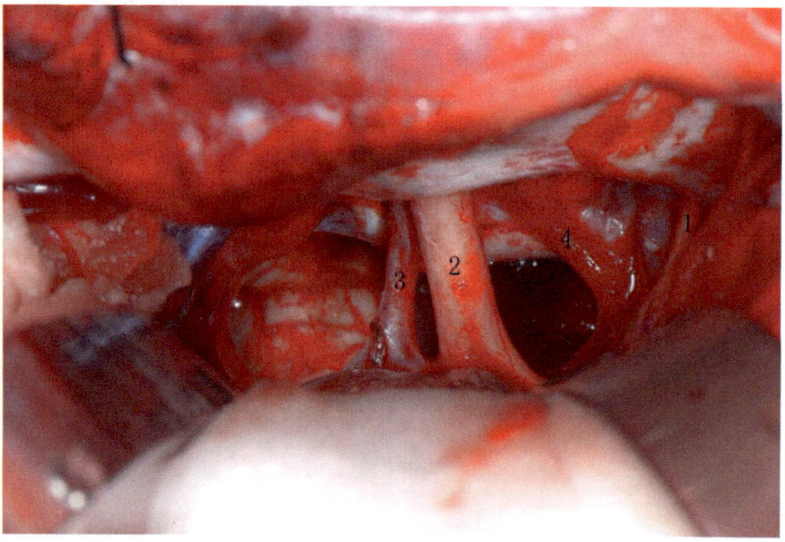

Fig. 32 The mass was removed totally. (*1*) contralateral olfactory tract, (*2*) optic nerve, (*3*) supraclinoidal segment of internal carotid artery, (*4*) pituitary stalk

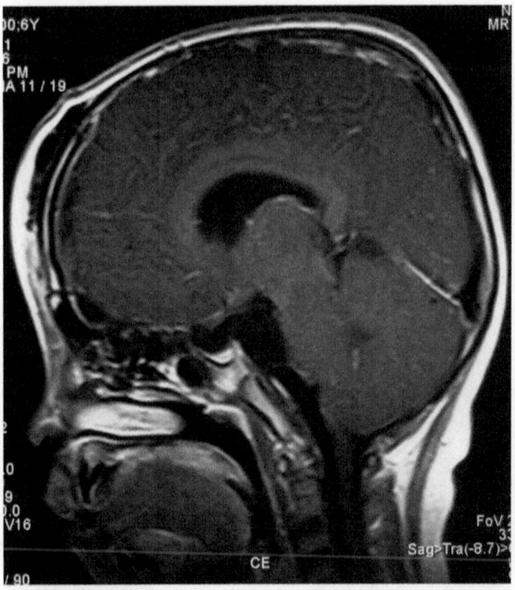

Fig. 33 Post-op sagittal Gd-enhanced MRI

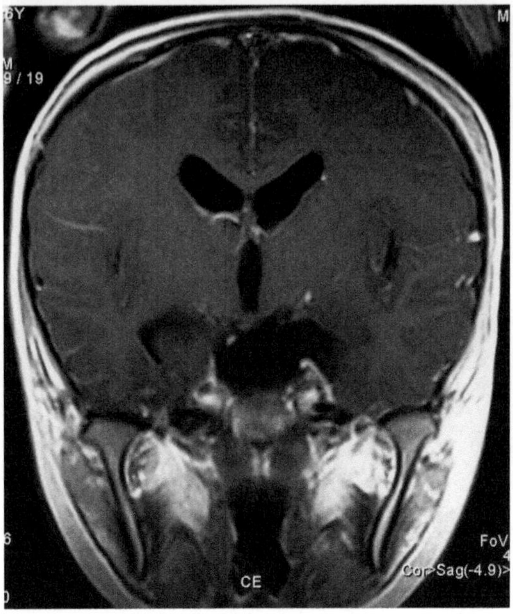

Fig. 34 Post-op coronal Gd-enhanced MRI

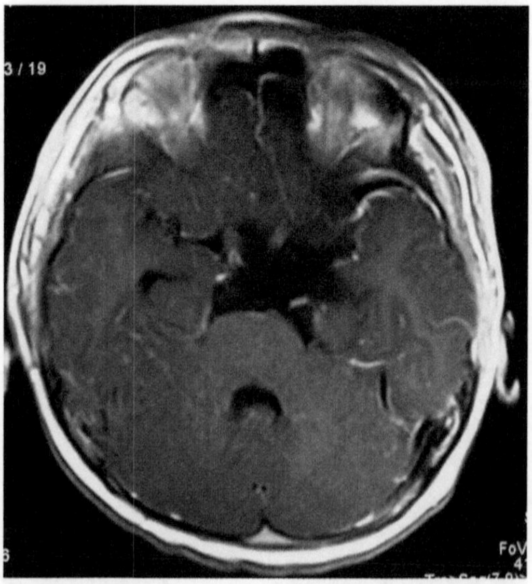

Fig. 35 Post-op axial Gd-enhanced MRI. Post-op Gd-enhanced MRI revealed gross resection of lesion

Case 2.2 Adult Craniopharyngioma (Frontobasal Approach)

Clinical presentation:

A 30-year-old female presented with a 1-month history of severe headache.

Neurological exam:

Decreased visual acuity (Lt: 0.8; Rt: 0.1), bitemporal hemianopsia

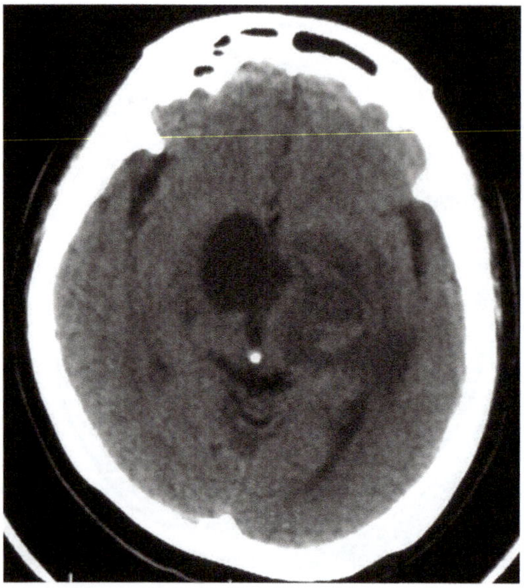

Fig. 36 Pre-op axial CT scan demonstrated a right suprasellar cystic mass with rare calcification

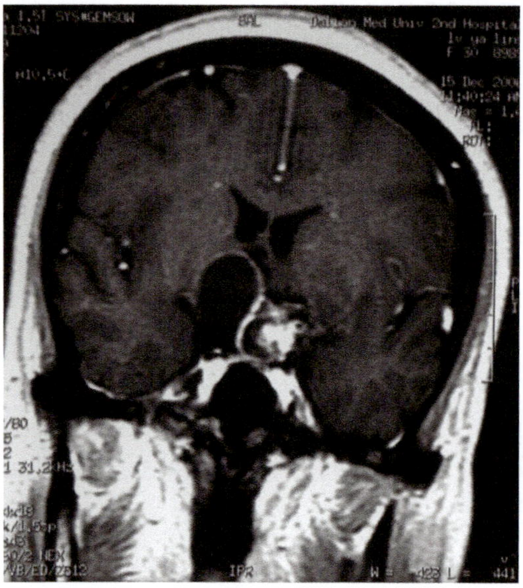

Fig. 37 Pre-op coronal T1 Gd-enhanced MRI

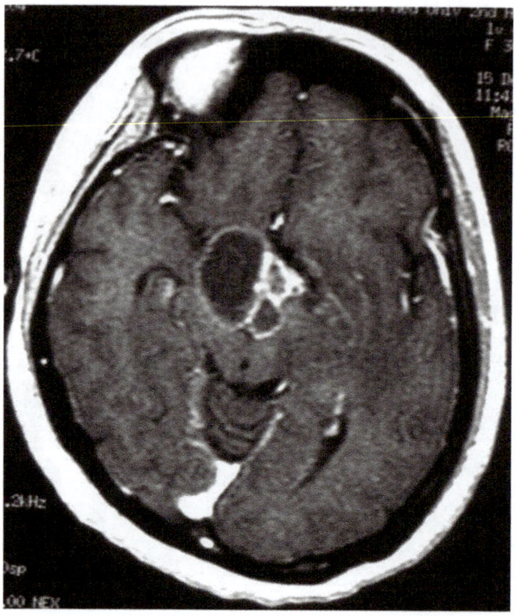

Fig. 38 Pre-op axial T1 Gd-enhanced MRI. Pre-op Gd-enhanced MRI showed significant enhancement of cystic wall and solid tumor components. The 3rd ventricle was distorted to the left side

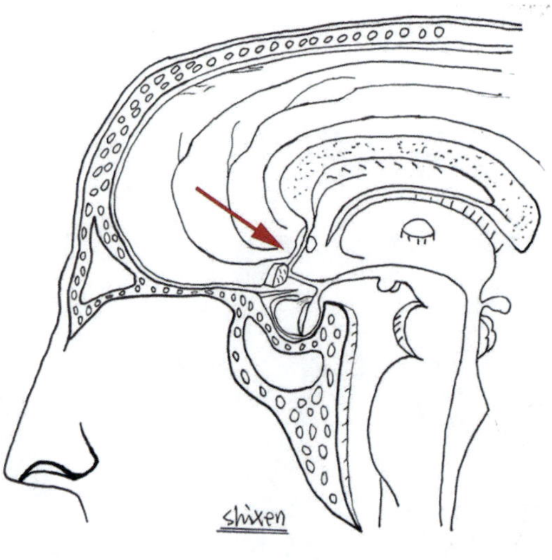

Fig. 39 The medical illustration revealed surgical trajectory of frontobasal interhemispheric approach to the suprasellar region and 3rd ventricle

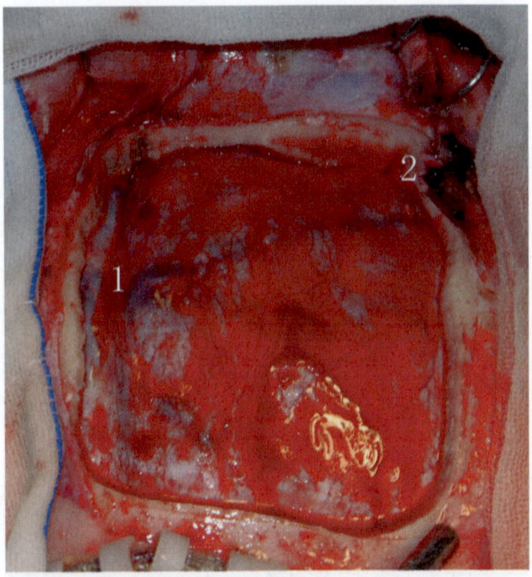

Fig. 40 The frontal musculofascial flap was retracted anteriorly. Right frontal dura was exposed following bone flap removal. (*1*) Superior sagittal sinus, (*2*) Keyhole at superior temporal line

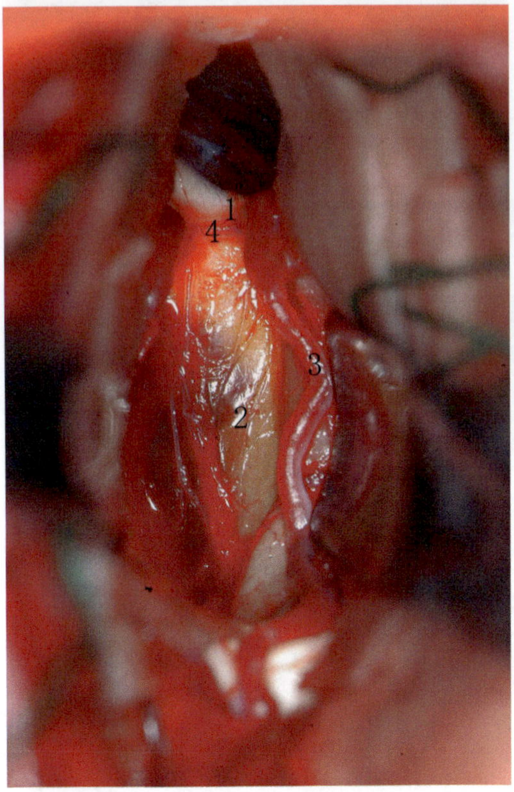

Fig. 41 The frontobasal interhemispheric fissure located anterior to the crus of corpus callosum was widely opened. (*1*) Optic chiasm, (*2*) Lamina terminalis, (*3*) Right A2, and (*4*) Anterior communicating artery

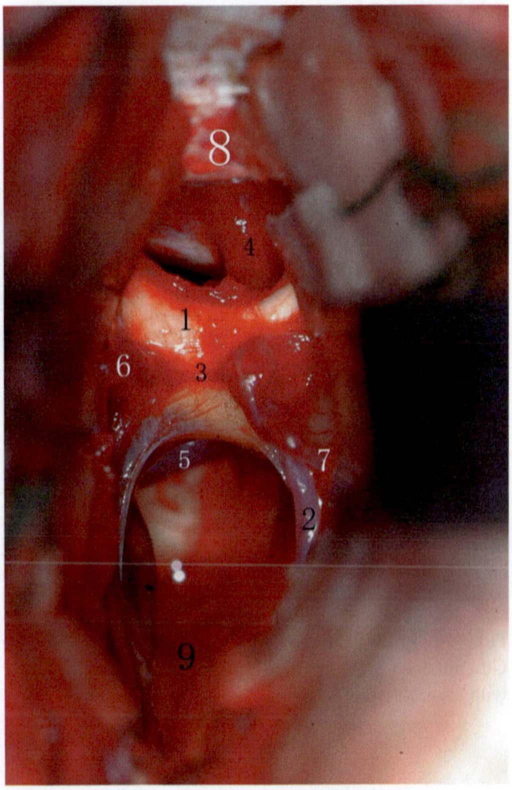

Fig. 42 After tumor resection, the following structures were unveiled. (*1*) Optic chiasm, (*2*) Right A2, (*3*) anterior communicating artery, (*4*) Pituitary stalk, (*5*) Left P1, (*6*) Left A2, (*7*) Right orbitofrontal artery, (*8*) Tuberculum sellae, and (*9*) Midbrain

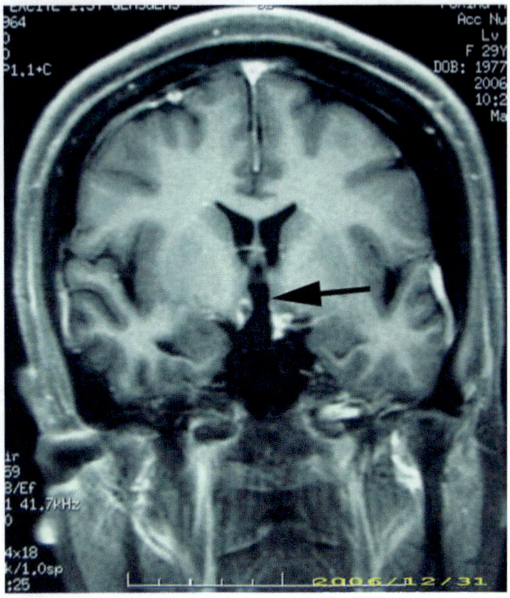

Fig. 43 Post-op coronal Gd-enhanced MRI; arrow indicated 3rd ventricle

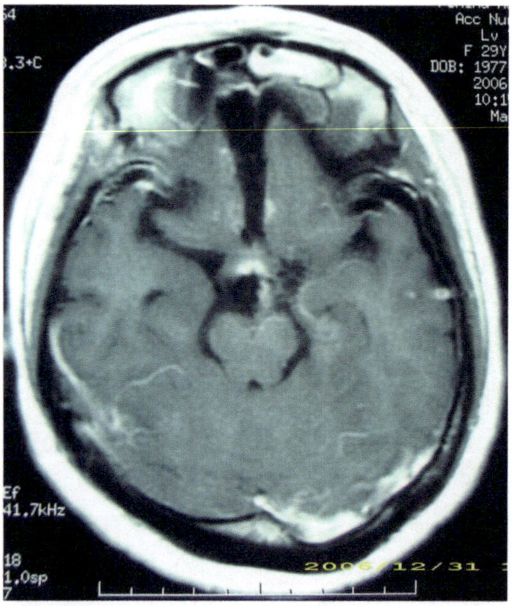

Fig. 44 Post-op axial Gd-enhanced MRI. Post-op Gd-enhanced MRI revealed gross resection of the tumor with favorable preservation of hypothalamus

Case 2.3 Multiple Anterior Circulation Aneurysms

Clinical presentation:

A 40-year-old lady presented with a 1-week history of sudden headache.

Neurological exam:

Nuchal rigidity

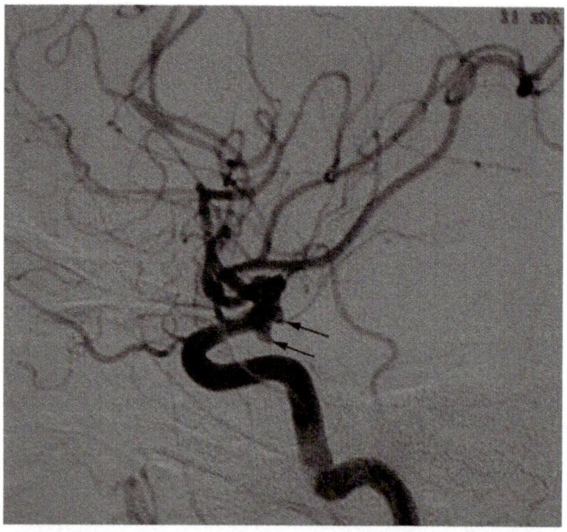

Fig. 45 Lateral view of left internal carotid artery angiography showed two saccular aneurysms. One involved posterior communicating segment and the other implicated anterior choroidal segment

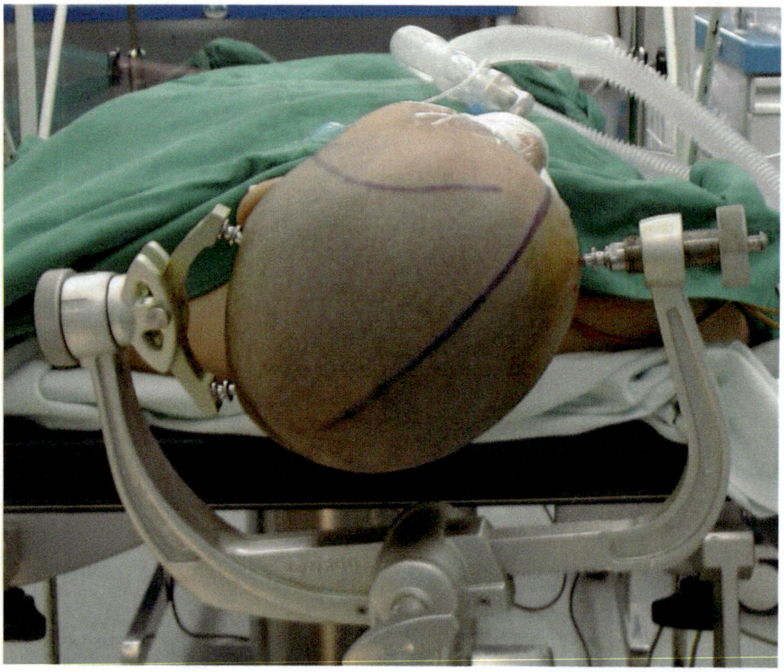

Fig. 46 Patient was placed in supine position with a 45° head rotation to the contralateral side

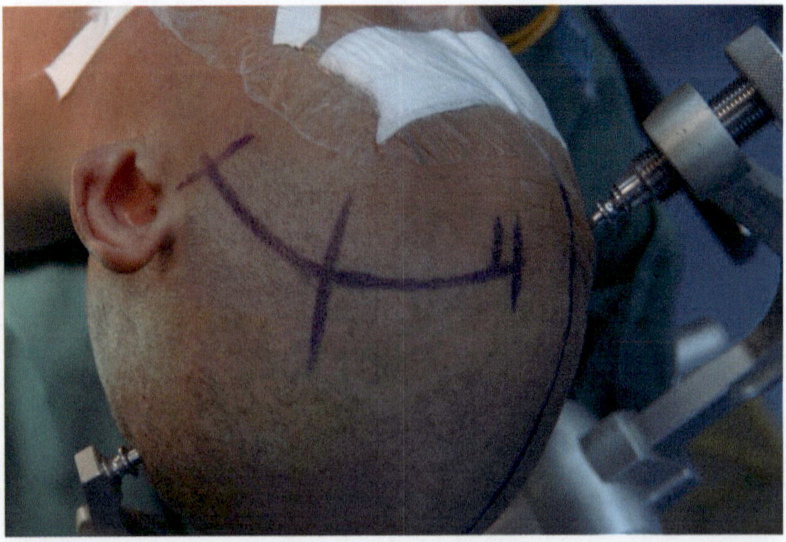

Fig. 47 The frontotemporal skin incision was marked

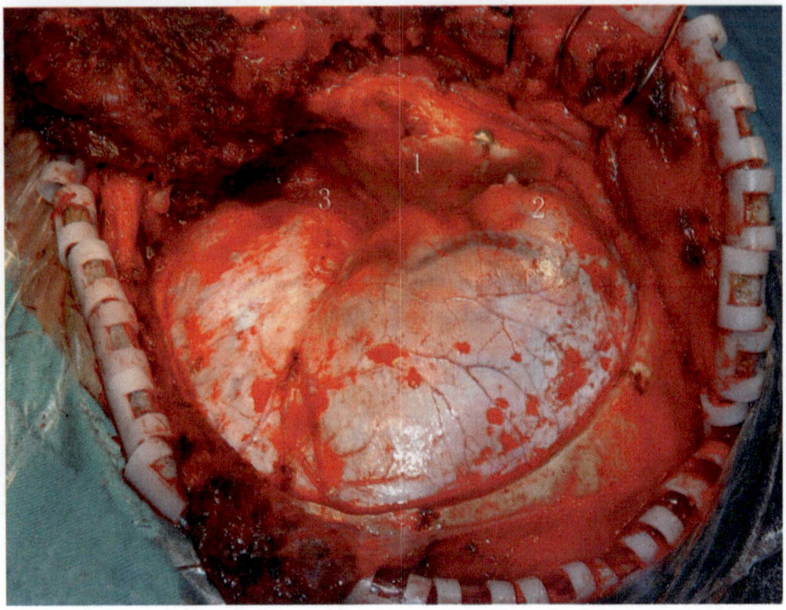

Fig. 48 Sphenoid ridge was flattened by cutting drill. (*1*) Sphenoid ridge, (*2*) Frontal dura, and (*3*) Temporal pole dura

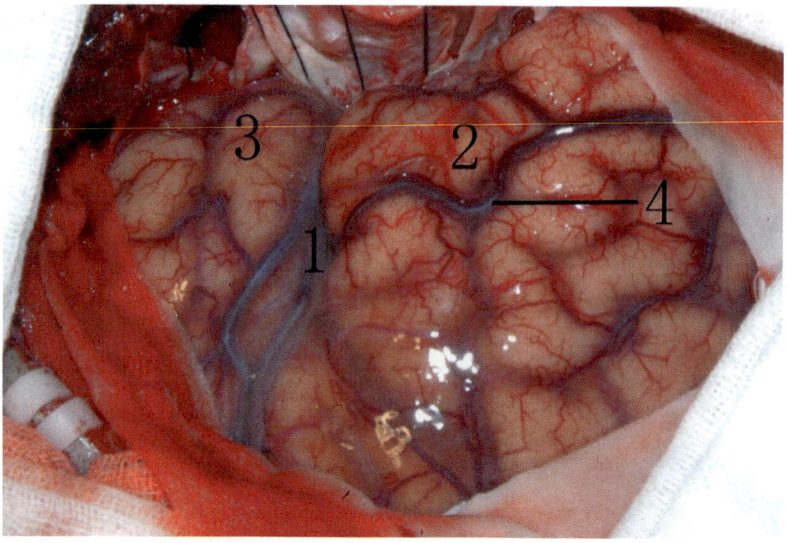

Fig. 49 Dura was opened and then perisylvian fissure region was exposed. (*1*) Sylvian fissure, (*2*) Pars orbitalis, (*3*) Temporal operculum, and (*4*) Frontosylvian vein

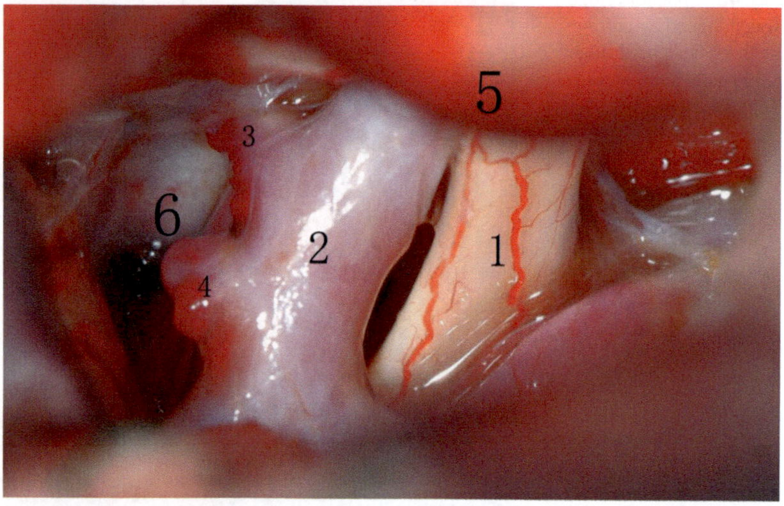

Fig. 50 After releasing the sylvian arachnoid, carotid and optic cistern was approached. (*1*) Left optic nerve, (*2*) Left supraclinoidal segment of internal carotid artery, (*3*) Posterior communicating segment aneurysm, (*4*) Anterior choroidal segment aneurysm, (*5*) Anterior clinoid process, and (*6*) Posterior clinoid process

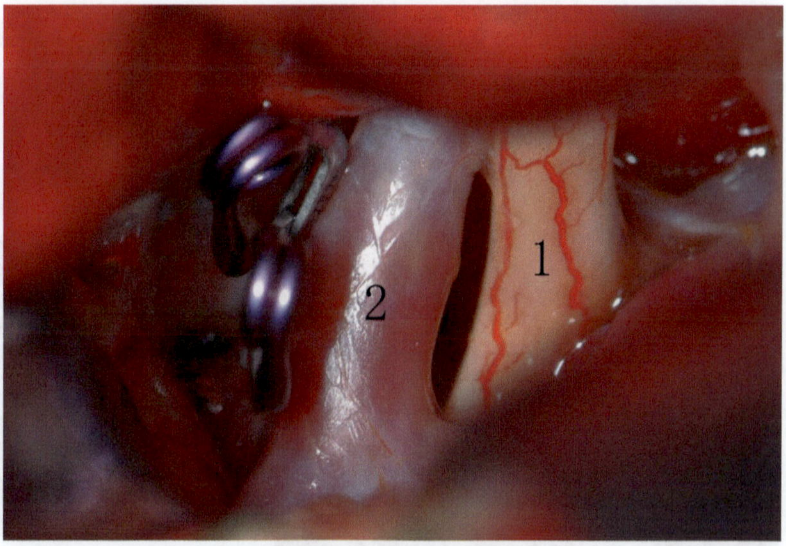

Fig. 51 Both aneurysms were repaired by a straight mini Yasagril clip. (*1*) Left optic nerve, (*2*) Left supraclinoid segment of internal carotid artery

Case 2.4 Craniopharyngioma (Pterional Approach)

Clinical presentation:

A 30-year-old lady presented with a 6-month history of polydipsia and polyuria.

Neurological exam:

Dry lips

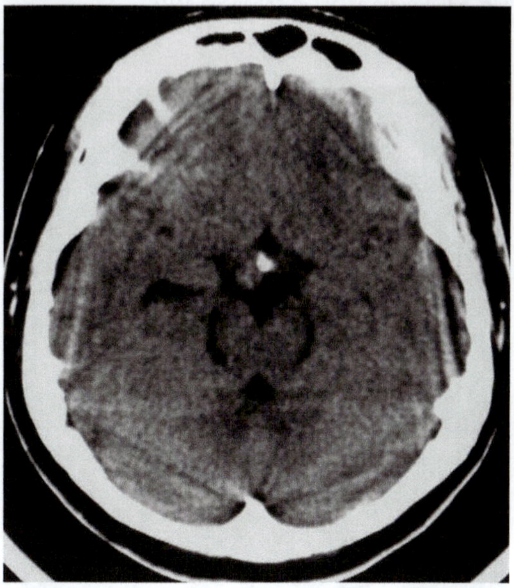

Fig. 52 Pre-op axial CT scan revealed a solid right suprasellar tumor with mixed intensity, indicating a calcified component

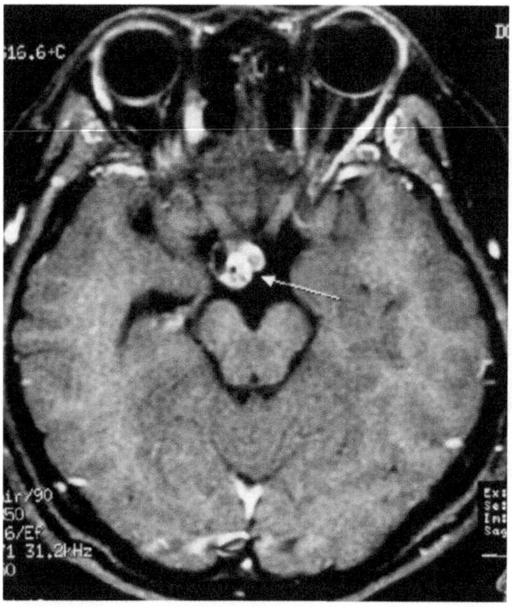

Fig. 53 Pre-op axial Gd-enhanced MRI

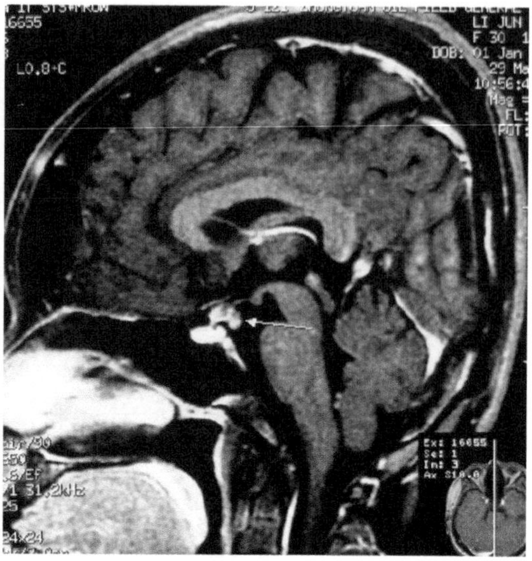

Fig. 55 Pre-op sagittal Gd-enhanced MRI. Pre-op Gd-enhanced MRI illustrated enhancement of cyst wall and solid tumor components

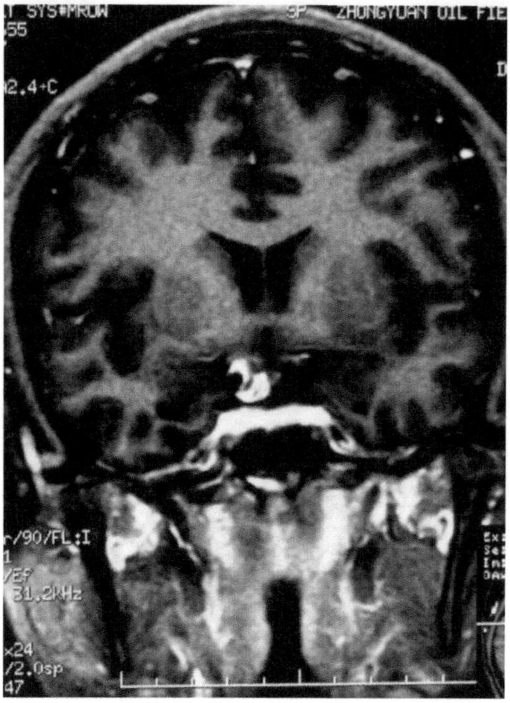

Fig. 54 Pre-op coronal Gd-enhanced MRI

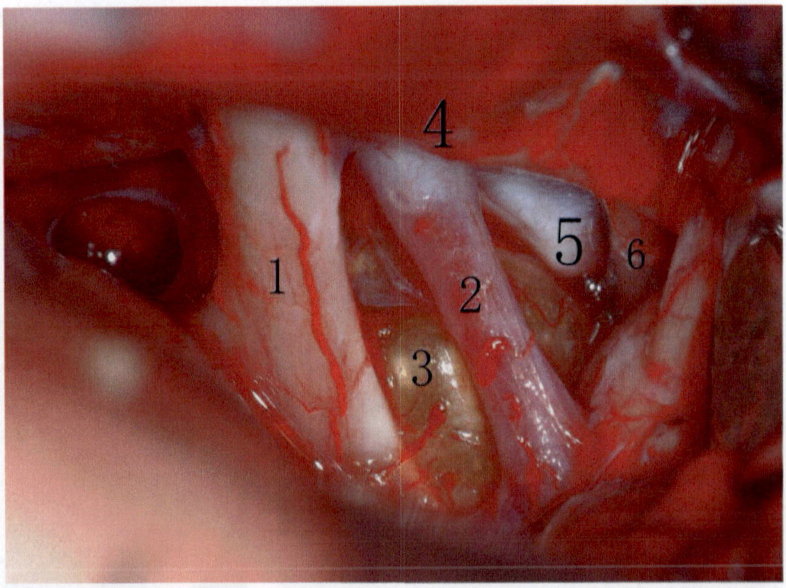

Fig. 56 The right pterional approach was chosen. The mass was identified and approached through optico-carotid and carotid-oculomotor triangle (space). (*1*) Right optic nerve, (*2*) right supraclinoid segment of internal carotid artery, (*3*) tumor, (*4*) anterior clinoid process, (*5*) posterior clinoid process, and (*6*) oculomotor nerve

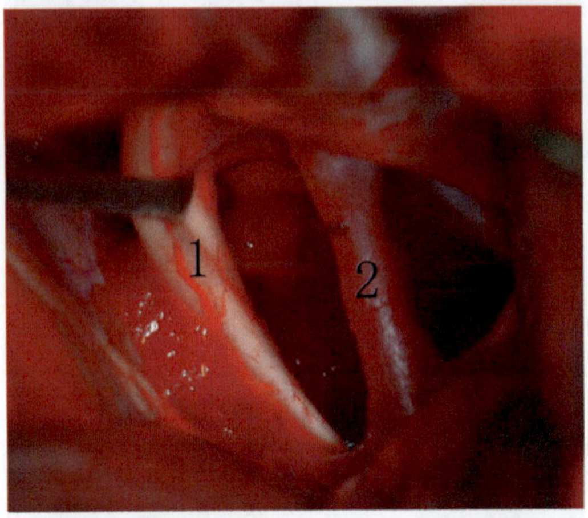

Fig. 57 Lesion was grossly removed. (*1*) Right optic nerve, (*2*) right supraclinoid segment of internal carotid artery

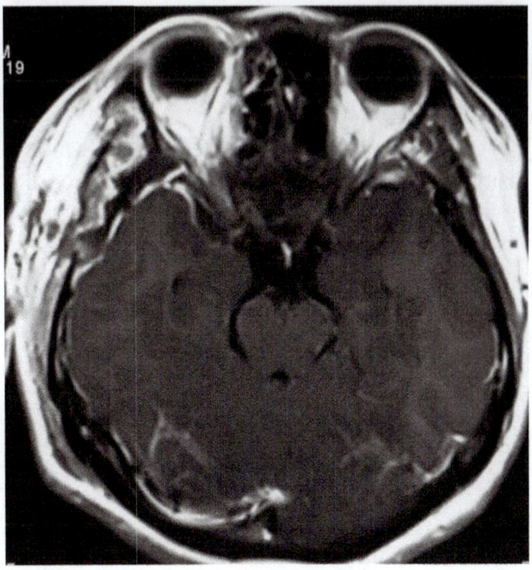

Fig. 58 Post-op axial Gd-enhanced MRI

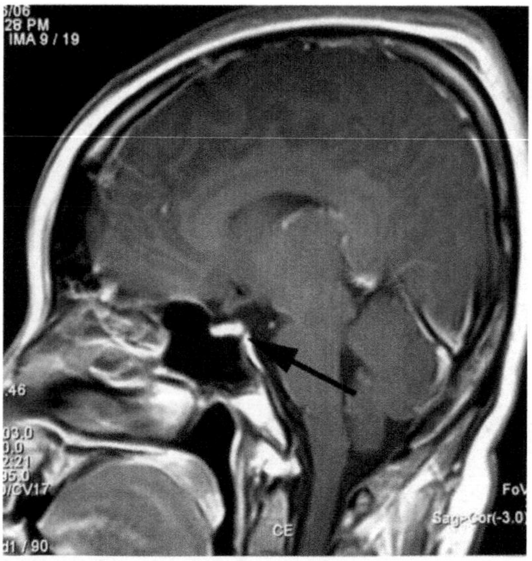

Fig. 59 Post-op sagittal Gd-enhanced MRI. Post-op Gd-enhanced MRI reveals disappearance of suprasellar mass

Case 2.5 Anterior Clinoidal Meningioma

Clinical presentation:

A 61-year-old male presented with a 2-month history of visual deficits.

Neurological exam:

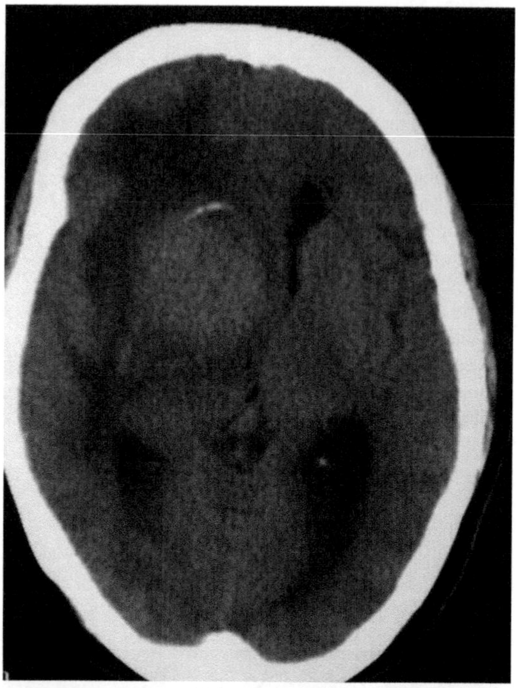

Fig. 60 Pre-op axial CT scan indicated a slight hyperintensity lesion of right frontal horn with massive perinodual edema

Unremarkable

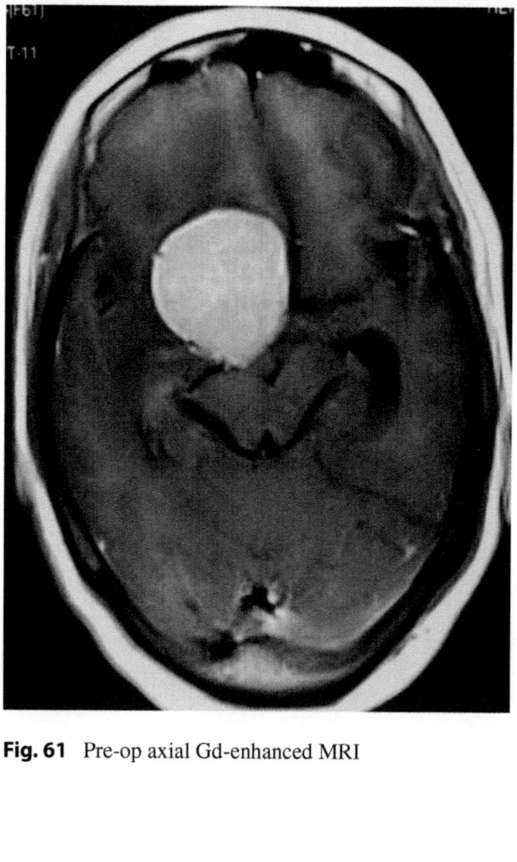

Fig. 61 Pre-op axial Gd-enhanced MRI

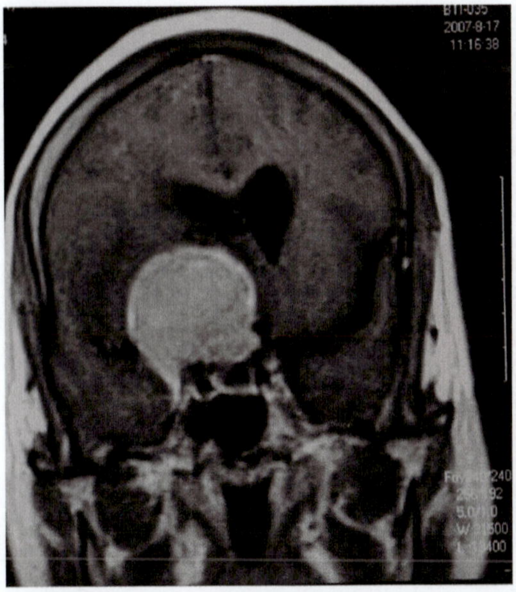

Fig. 62 Pre-op coronal Gd-enhanced MRI

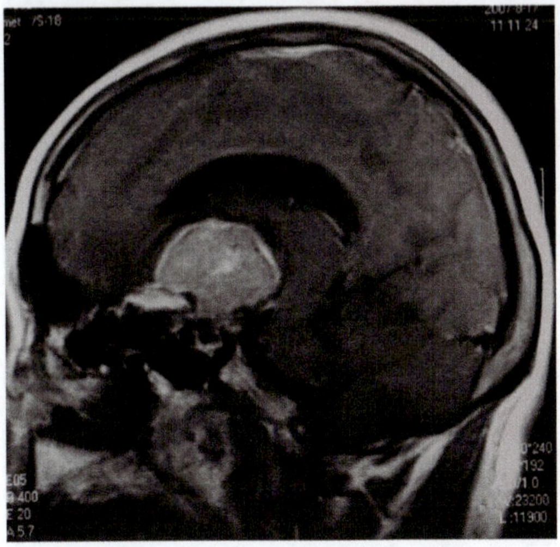

Fig. 63 Pre-op sagittal Gd-enhanced MRI. Pre-op Gd-enhanced MRI revealed a right parasellar lesion which demonstrated homogeneous significant enhancement with "dural tail" sign at anterior clinoid process. The frontal horn and 3rd ventricle were deformed

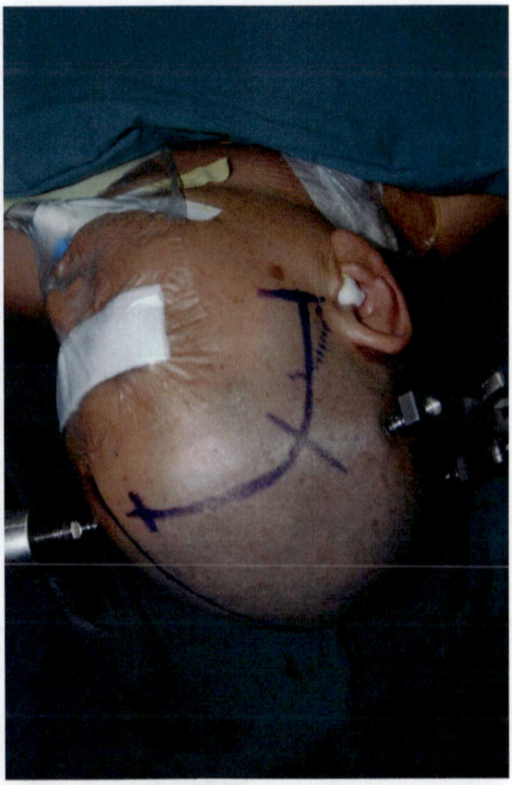

Fig. 64 The incision of right frontotemporal craniotomy and course of superficial temporal artery was labeled

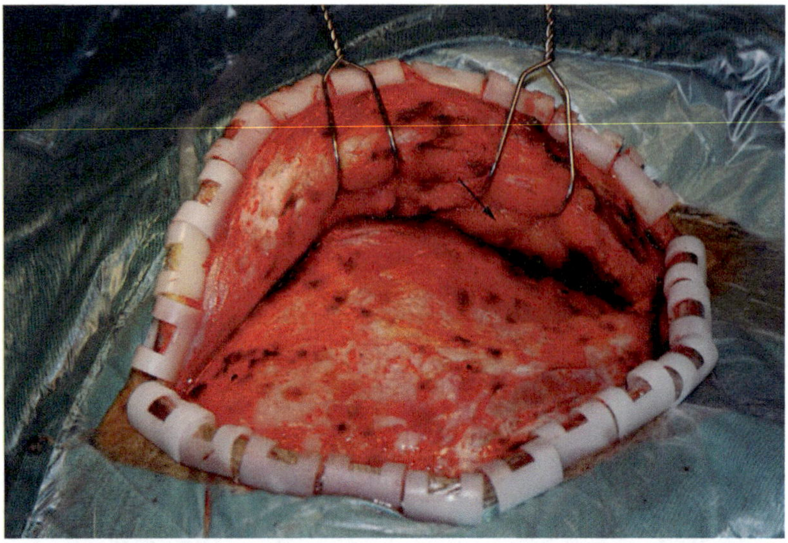

Fig. 65 The interfascial fat pad and its superficial tissues were reflected together to spare facial nerve branches

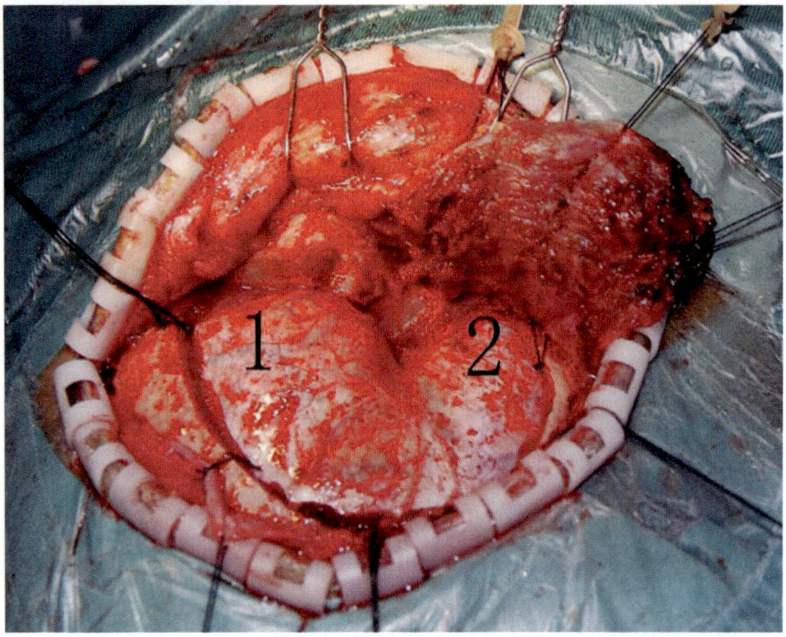

Fig. 66 The orbitomeningeal band was identified after flatting of sphenoid ridge. (*1*) Frontal dura and (*2*) temporal dura

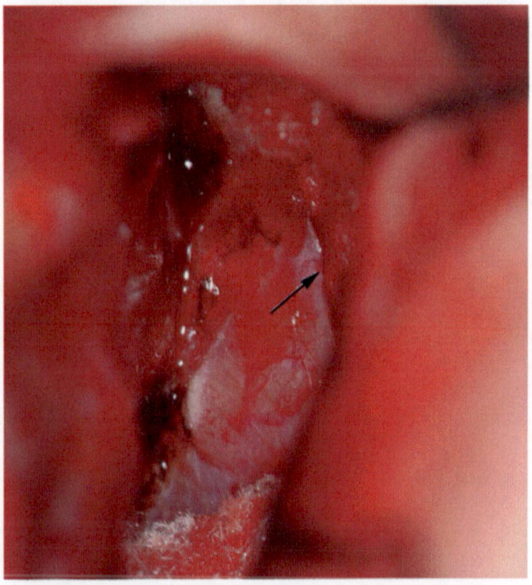

Fig. 67 With arachnoid dissection of Sylvian fissure, the base of lesion was located at the anterior clinoid process as indicated by the arrow

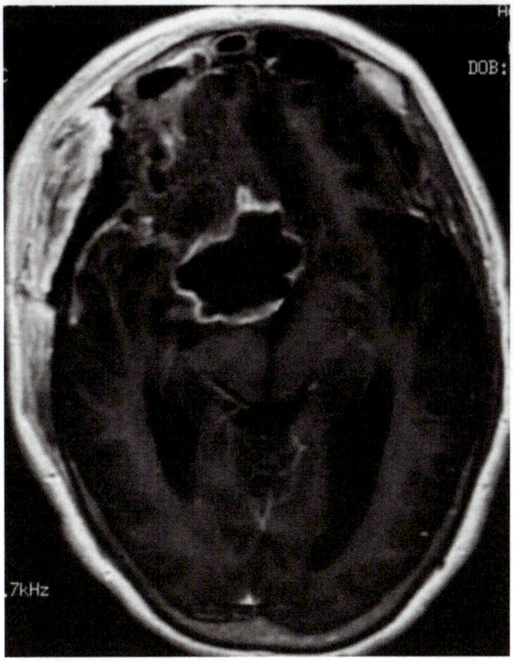

Fig. 68 Post-op axial Gd-enhanced MRI

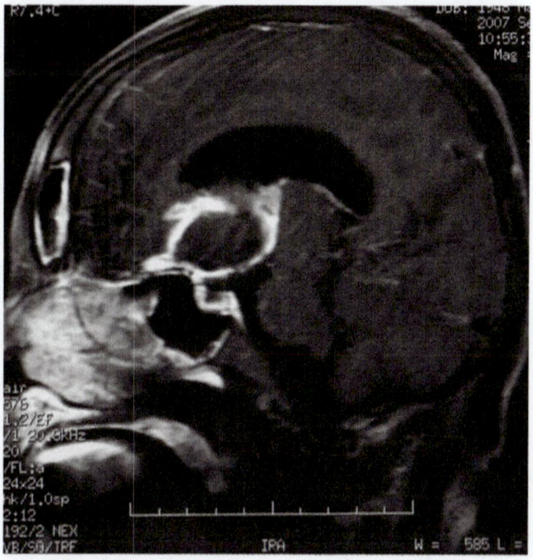

Fig. 69 Post-op sagittal Gd-enhanced MRI. Post-op Gd-enhanced MRI showed gross mass resection

Anatomy of Mesial Temporal Region, Lateral Ventricle, 3rd Ventricle and Surgical Techniques

Xiang'En Shi, Long Wang, and Hai Qian

1 Anatomy of Mesial Temporal Region

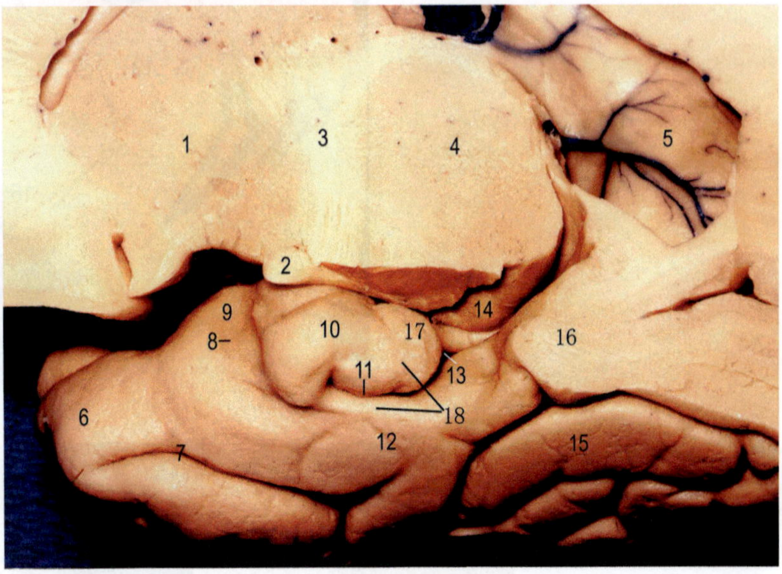

Fig. 1 Overview of right mesial temporal region. (*1*) Putamen, (*2*) Optic tract, (*3*) Internal capsule, (*4*) Thalamus, (*5*) Trigone of lateral ventricle, (*6*) Temporal pole, (*7*) Rhinal sulcus, (*8*) Amygdala, (*9*) Uncal apex, (*10*) Ambient gyrus, (*11*) Uncal notch (sulcus), (*12*) Parahippocampal gyrus, (*13*) Inferior choroidal point, (*14*) Gyrus fasciolaris, (*15*) Fusiform gyrus, (*16*) Lingual gyrus, (*17*) Band of Giacomini, (*18*) Posterior segment of the uncus

X. Shi (✉) · L. Wang · H. Qian
Neurological Surgery, SanBo Brain Hospital, Capital Medical University, Beijing, China
e-mail: shixen@ccmu.edu.cn; wanglong@ccmu.edu.cn; qianh@ccmu.edu.cn

X. Shi et al. (eds.), *Microneurosurgical Anatomy and Surgical Technique*,
https://doi.org/10.1007/978-981-19-8273-6_3

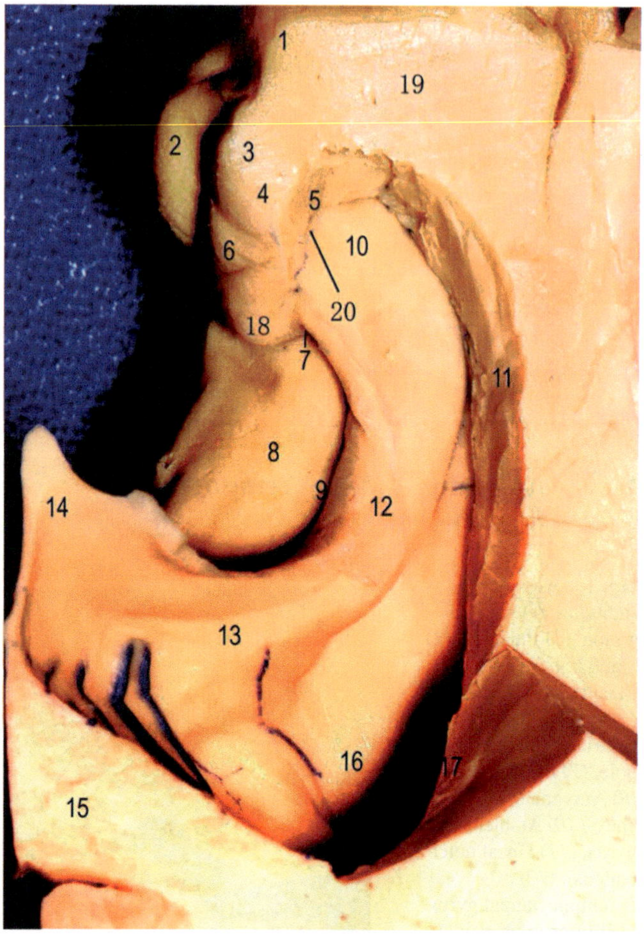

Fig. 2 Superior view of right mesial temporal region. (*1*) Anterior segment of uncus, (*2*) Parahippocampal gyrus, (*3*) Semilunar gyrus, (*4*) Ambient gyrus, (*5*) Amygdala, (*6*) Uncinate gyrus, (*7*) Inferior choroidal point, (*8*) Parahippocampal gyrus, (*9*) Dentate gyrus, (*10*) Head of hippocampus, (*11*) Tapetum, (*12*) Fimbria of fornix, (*13*) Tail of hippocampus, (*14*) Crus of fornix, (*15*) Splenium of corpus callosum, (*16*) Collateral trigone, (*17*) Roof of trigone part of lateral ventricle, (*18*) Intralimbic gyrus, (*19*) Lentiform nucleus, (*20*) Uncal recess

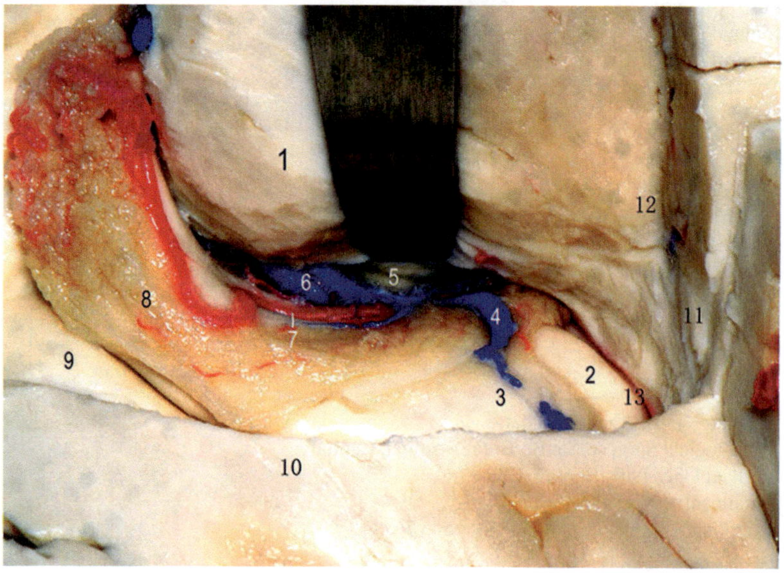

Fig. 3 Lateral view of right mesial temporal region. (*1*) Thalamus, (2) Head of hippocampus, (*3*) Body of hippocampus, (*4*) Inferior ventricular vein, (*5*) Lateral geniculate body, (*6*) Basal vein of Rosenthal, (*7*) Anterior choroidal artery, (*8*) Choroid plexus, (*9*) Calcar avis, (*10*) Middle temporal gyrus, (*11*) Amygdala, (*12*) Lentiform nucleus, (*13*) Uncal recess

2 Anatomy of Lateral Ventricle

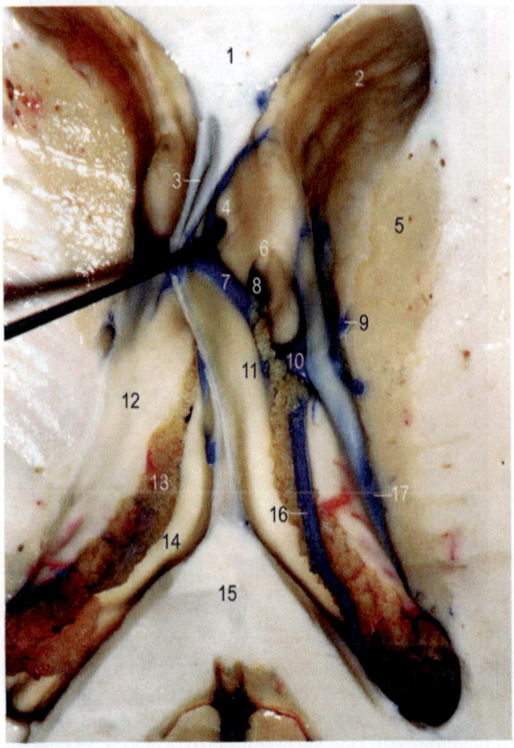

Fig. 4 Overview of lateral ventricle. (*1*) Genu of corpus callosum, (*2*) Frontal horn, (*3*) Cavum septum pellucidum, (*4*) Left column of fornix, (*5*) Head of caudate nucleus, (*6*) Right column of fornix, (*7*) Septal vein, (*8*) Foramen of Monro, (*9*) Middle caudate vein, (*10*) Thalamostriate vein, (*11*) Body of fornix, (*12*) Thalamostriate groove, (*13*) Choroid plexus, (*14*) Crus of fornix, (*15*) Splenium of corpus callosum, (*16*) Superior choroidal vein, (*17*) Thalamostriate vein

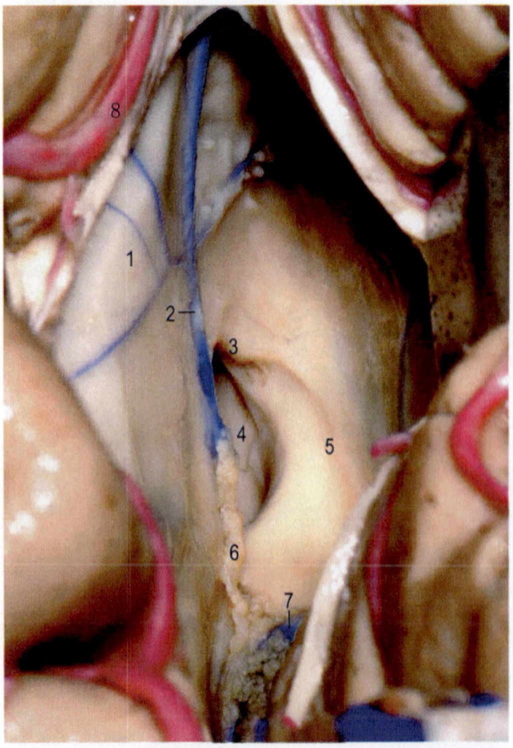

Fig. 5 Overview of the frontal horn of right lateral ventricle. (*1*) Septum pellucidum, (*2*) Septal vein, (*3*) Column of fornix, (*4*) Foramen of Monro, (*5*) Thalamostriate groove, (*6*) Choroid plexus, (*7*) Superior choroidal vein, (*8*) Anterior cerebral artery

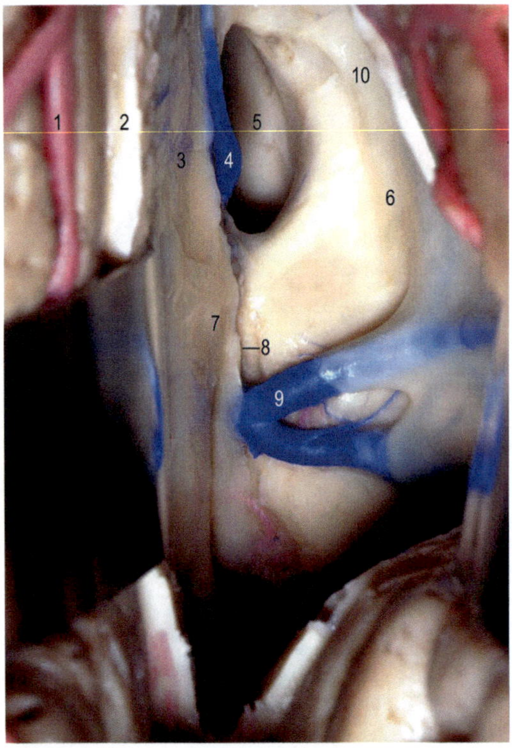

Fig. 6 Overview of the body of right lateral ventricle. (*1*) Anterior cerebral artery, (*2*) Corpus callosum, (*3*) Septum pellucidum, (*4*) Septal vein, (*5*) Foramen of Monro, (*6*) Thalamostriate groove, (*7*) Body of fornix, (*8*) Choroidal fissure, (*9*) Thalamostriate vein, (*10*) Head of caudate nucleus

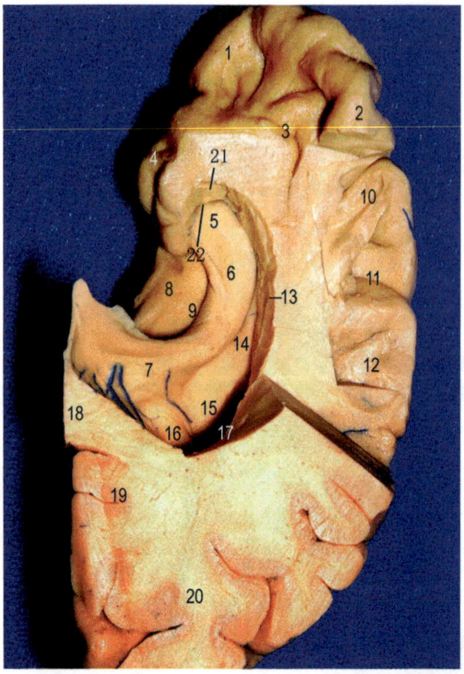

Fig. 7 Overview of the temporal horn of right lateral ventricle. (*1*) Planum Polare, (*2*) Superior temporal gyrus, (*3*) Limen insula, (*4*) Uncus, (*5*) Head of hippocampus, (*6*) Body of hippocampus, (*7*) Tail of hippocampus, (*8*) Subiculum, (*9*) Dentate gyrus, (*10*) Anterior transverse temporal gyrus, (*11*) Middle transverse temporal gyrus, (*12*) Posterior transverse temporal gyrus, (*13*) Tapetum, (*14*) Anterior part of collateral trigone, (*15*) Posterior part of collateral trigone, (*16*) Calcar avis, (*17*) Roof of trigone of lateral ventricle, (*18*) Splenium of corpus callosum, (*19*) Precuneus, (*20*) Cuneus, (*21*) Amygdala, (*22*) Uncal recess

Fig. 8 Overview of the trigone and occipital horn of right lateral ventricle. (*1*) Superior sagittal sinus, (*2*) Occipital vein, (*3*) Crus of fornix, (*4*) Bulb of corpus callosum, (*5*) Tela choroidea, (*6*) Choroidal fissure, (*7*) Pulvinar, (*8*) Calcar avis, (*9*) Occipital horn, (*10*) Transverse sinus

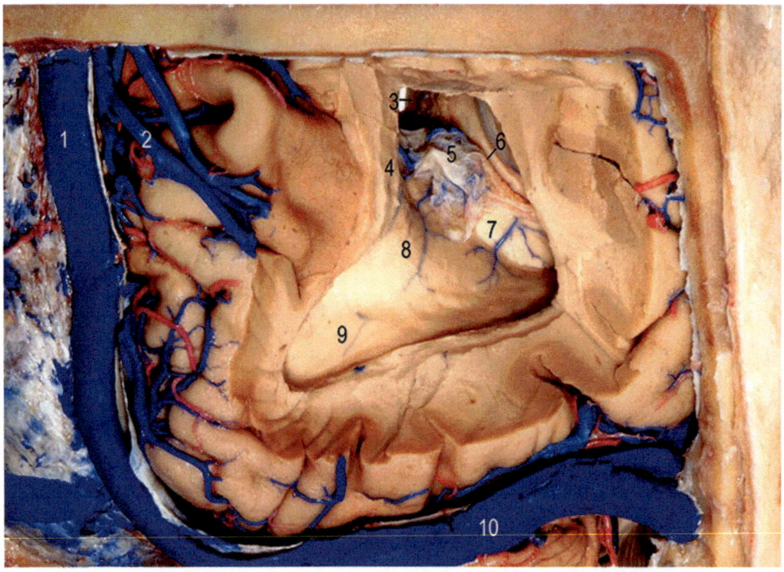

3 Anatomy of 3rd Ventricle

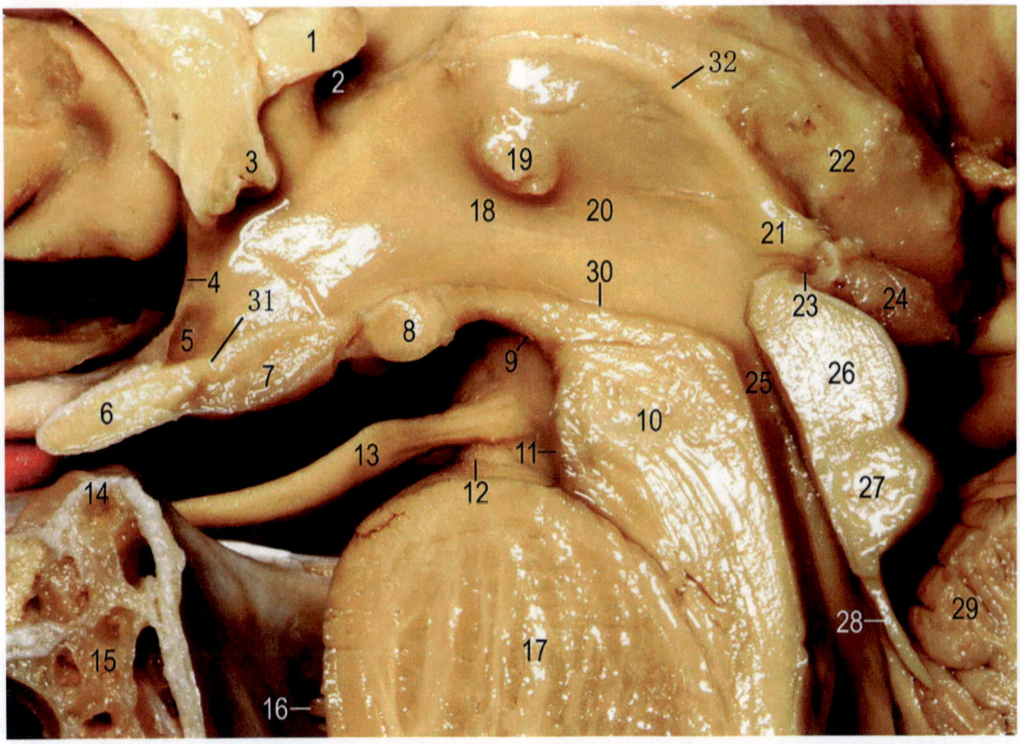

Fig. 9 Sagittal plane of 3rd ventricle. (*1*) Column of fornix, (*2*) Foramen of Monro, (*3*) Anterior commissure, (*4*) Lamina terminalis, (*5*) Optic recess, (*6*) Optic chiasm, (*7*) Tuber cinereum, (*8*) Mamillary body, (*9*) Posterior perforated substance, (*10*) Pons, (*11*) Interpeduncular fossa, (*12*) Pontomesencephalic sulcus, (*13*) Oculomotor nerve, (*14*) Dorsum sellae, (*15*) Clivus, (*16*) Trigeminal nerve, (*17*) Pons, (*18*) Hypothalamic sulcus, (*19*) Massa intermedia, (*20*) Thalamus, (*21*) Habenular commissure, (*22*) Hippocampal commissure, (*23*) Pineal recess, (*24*) Pineal gland, (*25*) Cerebral aqueduct, (*26*) Superior colliculus, (*27*) Inferior colliculus, (*28*) Superior medullary velum, (*29*) Lingula, (*30*) Thalamomesencephalon sulcus, (*31*) Median eminence, (*32*) Stria medullaris thalami

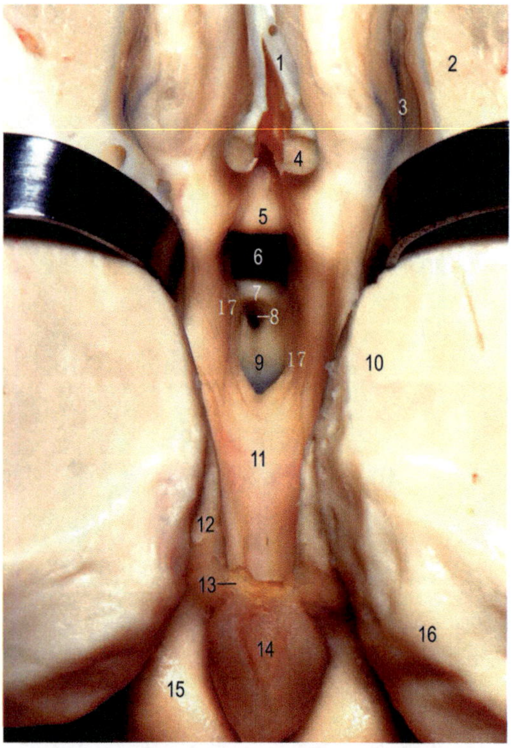

Fig. 10 Overview of the 3rd ventricle. (*1*) Septum pellucidum, (*2*) Head of caudate nucleus, (*3*) Anterior caudate vein, (*4*) Column of fornix, (*5*) Anterior commissure, (*6*) Optic recess, (*7*) Optic chiasm, (*8*) Infundibular recess, (*9*) Tuber cinereum, (*10*) Medial part of thalamus, (*11*) Midbrain, (*12*) Habenular trigone, (*13*) Habenular commissure, (*14*) Pineal gland, (*15*) Superior colliculus, (*16*) Pulvinar (ventricular part), (*17*) Hypothalamic sulcus

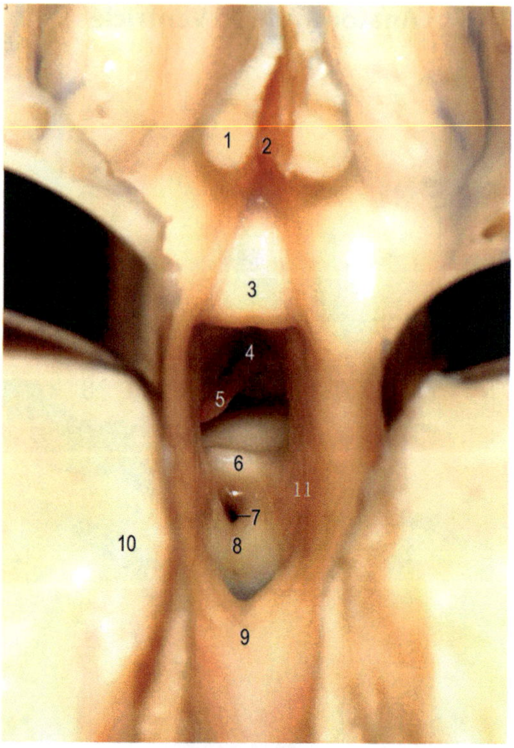

Fig. 11 Floor of 3rd ventricle. (*1*) Column of fornix, (*2*) Commissure of fornix, (*3*) Anterior commissure, (*4*) Lamina terminalis, (*5*) Anterior cerebral artery, (*6*) Optic chiasm, (*7*) Infundibular recess, (*8*) Tuber cinereum, (*9*) Midbrain, (*10*) Thalamus, (*11*) Hypothalamus

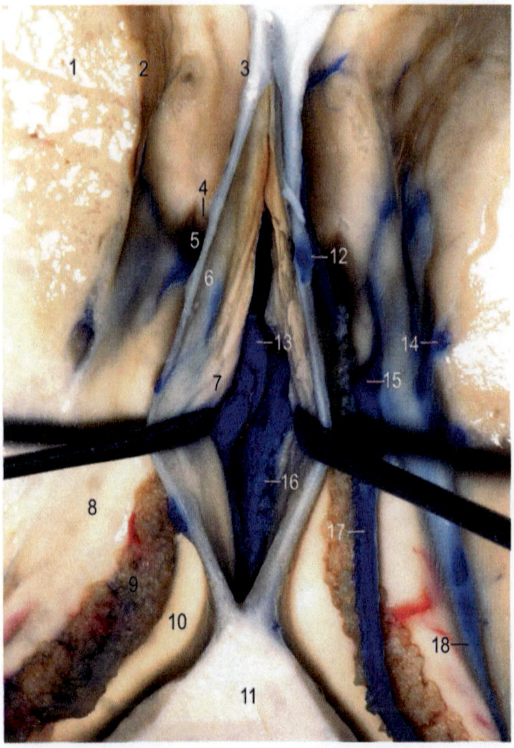

Fig. 12 Opening the roof of 3rd ventricle. (*1*) Head of caudate nucleus, (*2*) Lateral wall of frontal horn, (*3*) Medial wall of frontal horn, (*4*) Column of fornix, (*5*) Foramen of Monro, (*6*) Septum pellucidum, (*7*) Body of fornix, (*8*) Thalamostriate groove, (*9*) Choroid plexus, (*10*) Crus of fornix, (*11*) Splenium of corpus callosum, (*12*) Septal vein, (*13*) Left internal cerebral vein, (*14*) Middle caudate vein, (*15*) Thalamostriate vein, (*16*) Right internal cerebral vein, (*17*) Superior choroidal vein, (*18*) Thalamostriate vein

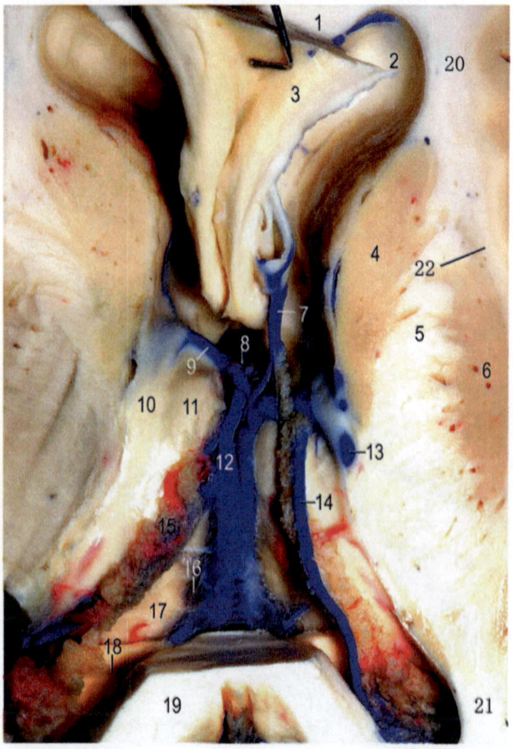

Fig. 13 Superficial layers of the roof of 3rd ventricle. (*1*) Genu of corpus callosum, (*2*) Frontal horn, (*3*) Body of fornix, (*4*) Head of caudate nucleus, (*5*) Internal capsule, (*6*) Lenticular nucleus, (*7*) Septal vein, (*8*) Foramen of Monro, (*9*) Thalamostriate vein, (10) Thalamostriate groove, (*11*) Thalamus, (*12*) Internal cerebral vein, (*13*) Middle caudate vein, (*14*) Superior choroidal vein, (*15*) Choroid plexus, (*16*) Superior layer of tela choroidea, (*17*) Pulvinar (ventricular part), (*18*) Crus of fornix, (*19*) Splenium of corpus callosum, (*20*) Genu of corpus callosum, (*21*) Tapetum, (*22*) External capsule

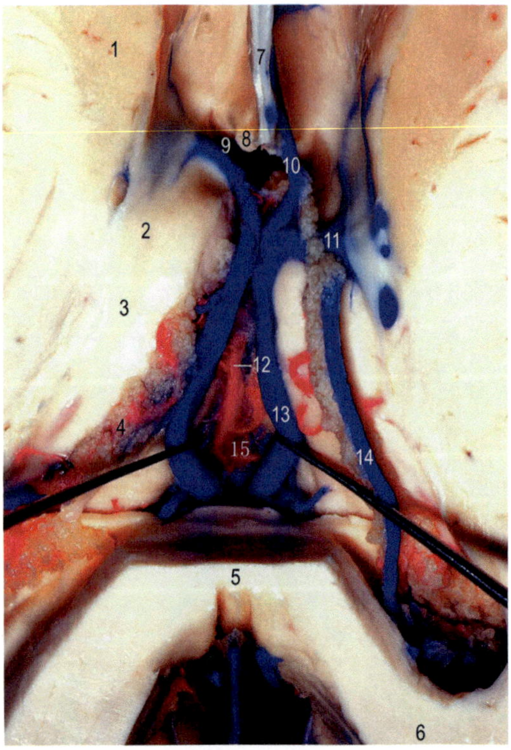

Fig. 14 Deep layers of the roof of 3rd ventricle. (*1*) Head of caudate nucleus, (*2*) Thalamostriate sulcus, (*3*) Thalamus, (*4*) Choroid plexus, (*5*) Splenium of corpus callosum, (*6*) Bulb of corpus callosum, (*7*) Septum pellucidum, (*8*) Column of fornix, (*9*) Foramen of Monro, (*10*) Septal vein, (*11*) Thalamostriate vein, (*12*) Medial posterior choroidal artery, (*13*) Internal cerebral vein, (*14*) Superior choroidal vein, (*15*) Inferior layer of tela choroidea

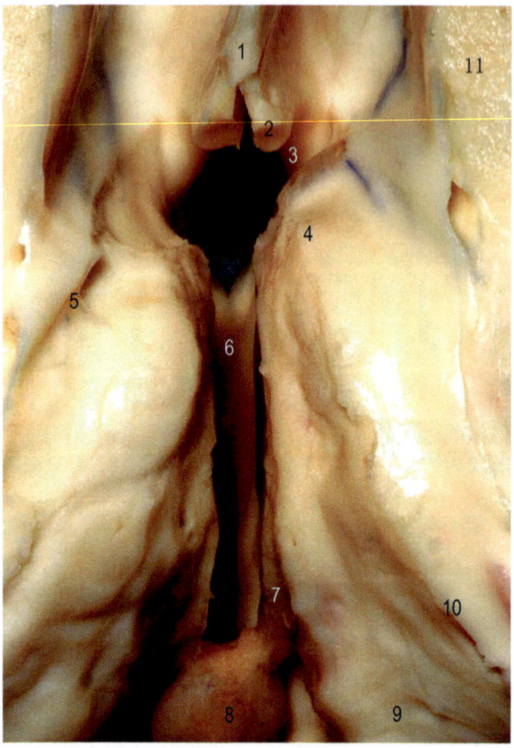

Fig. 15 Lateral wall of 3rd ventricle. (*1*) Septum pellucidum, (*2*) Column of fornix, (*3*) Foramen of Monro, (*4*) Anterior thalamic tubercle, (*5*) Thalamostriate groove, (*6*) Third ventricle, (*7*) Habenular trigone, (*8*) Pineal gland, (*9*) Pulvinar (ventricular part), (*10*) Thalamostriate groove, (*11*) Head of caudate nucleus

Case 3.1 3rd Ventricular Glioma

Clinical presentation:

A 62-year-old male presented with a 2-month history of headache and vomiting.

Neurological exam:

Unremarkable

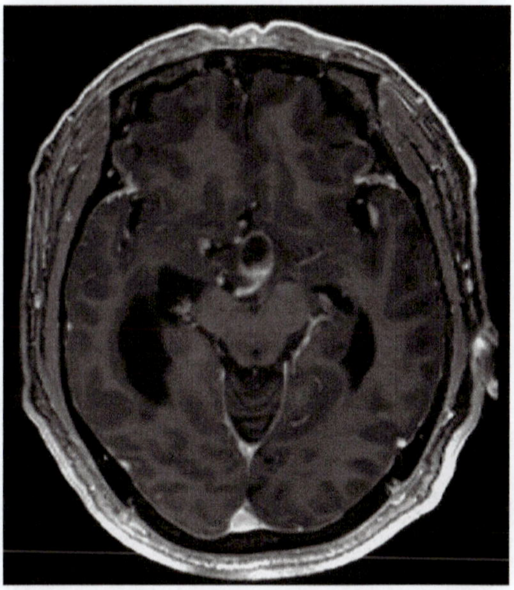

Fig. 16 Pre-op axial Gd-enhanced T1-weighted MRI

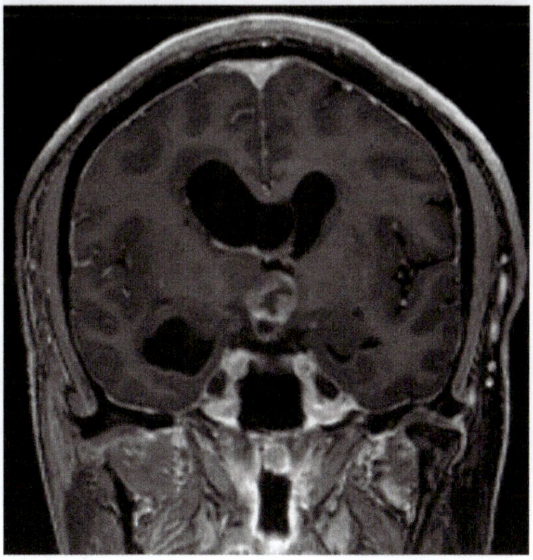

Fig. 17 Pre-op coronal Gd-enhanced T1-weighted MRI

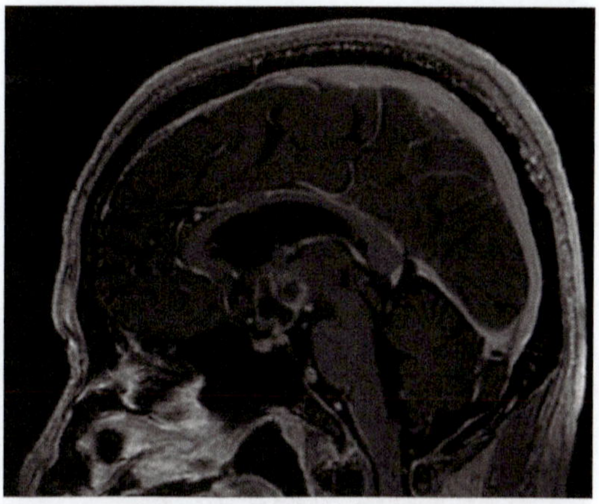

Fig. 18 Pre-op sagittal Gd-enhanced T1-weighted MRI. Pre-op Gd-enhanced T1-weighted MRI demonstrated a poorly defined cystic mass of 3rd ventricle with heterogeneous enhancement and dilation of lateral ventricles

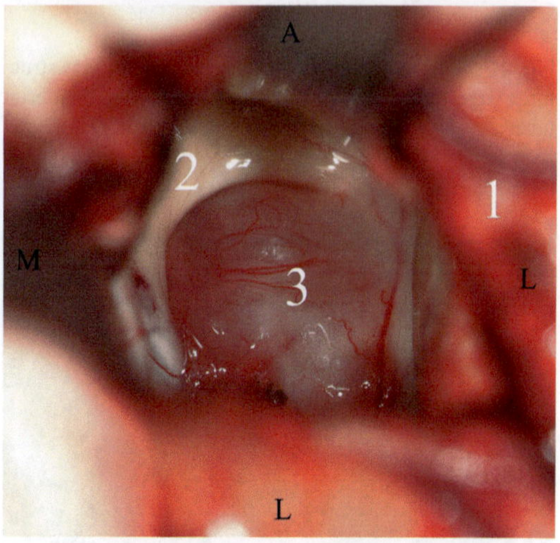

Fig. 19 The right frontal craniotomy and transcortical-transchoroidal fissure approach was performed to reach the nidus. (*1*) right middle frontal gyrus, (*2*) body of fornix, and (*3*) tumor

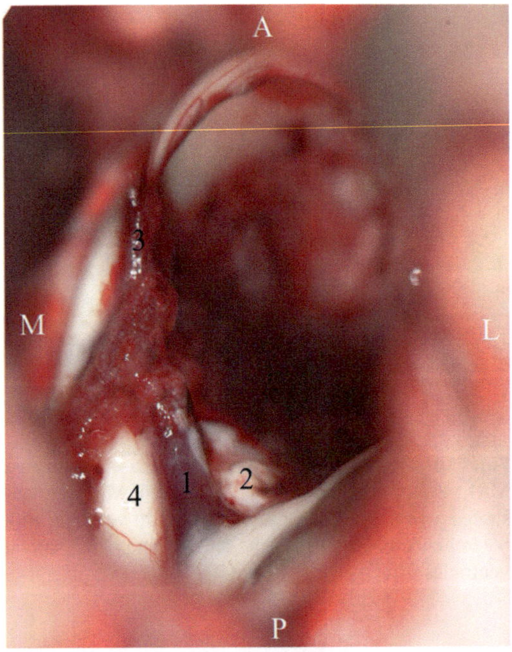

Fig. 20 After tumor resection, the intraventricular structures were identified. (*1*) thalamostriate vein, (*2*) anterior tubercle of thalamus, (*3*) right septal vein, (*4*) body of fornix

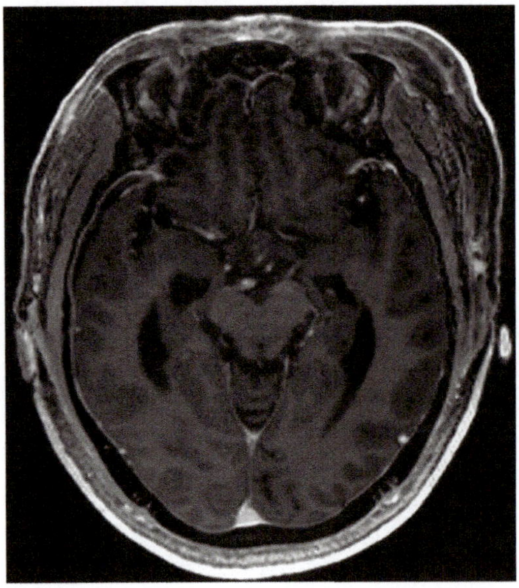

Fig. 21 Post-op axial Gd-enhanced T1-weighted MRI

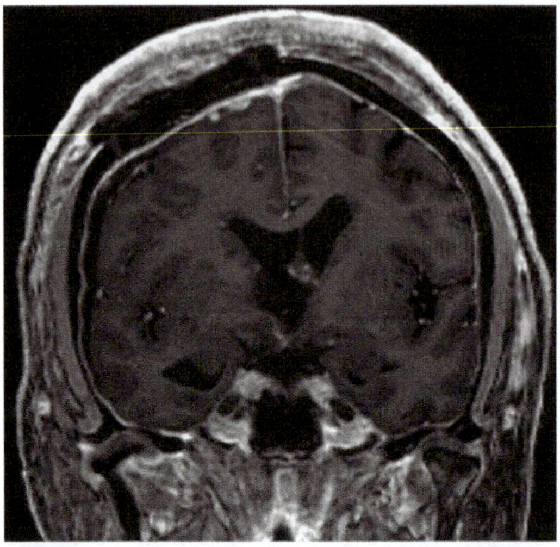

Fig. 22 Post-op coronal Gd-enhanced T1-weighted MRI

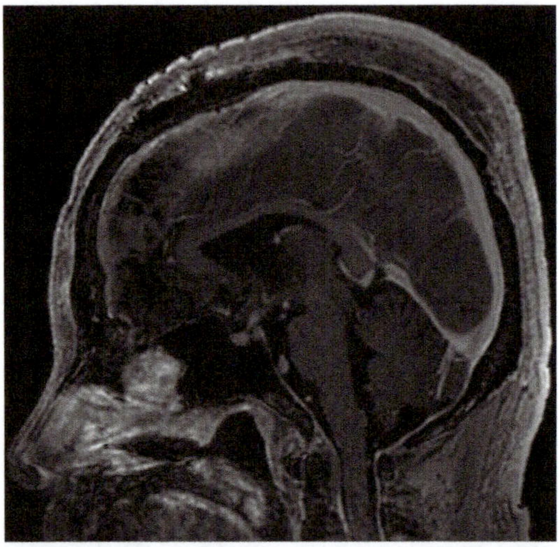

Fig. 23 Post-op sagittal Gd-enhanced T1-weighted MRI. Post-op Gd-enhanced T1-weighted MRI showed total resection of the mass

Case 3.2 Midbrain Cavernous Malformation

Clinical presentation:

A 58-year-old male presented with a 1-week history of hemiparesis of left upper extremity and hallucination.

Neurological exam:

Muscular strength: Grade III (left upper extremity)

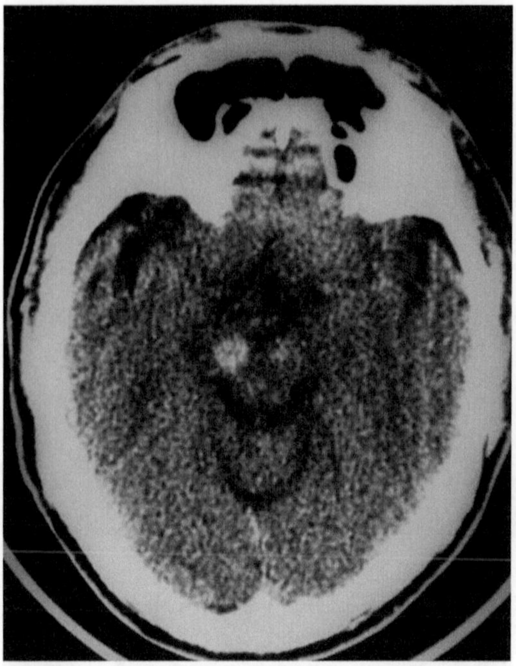

Fig. 24 Pre-op CT scan revealed a well-circumscribed hyperdense region at right cerebral peduncle, likely hemorrhage

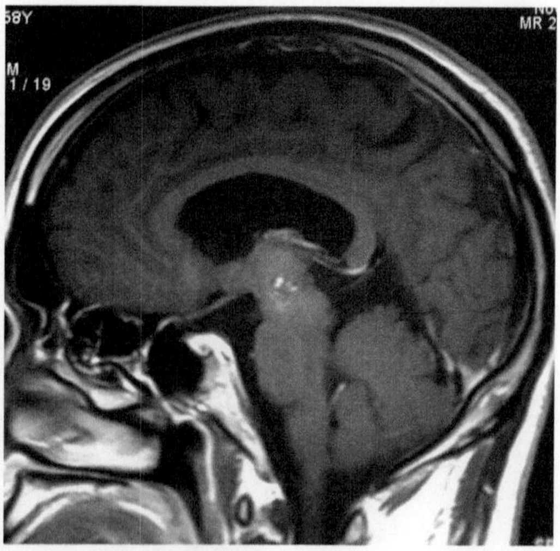

Fig. 26 Pre-op sagittal T1 contrast-enhanced MRI. Pre-op T1 contrast-enhanced MRI demonstrated a popcorn-like mass at midbrain without enhancement

Fig. 25 Pre-op axial T1 contrast-enhanced MRI

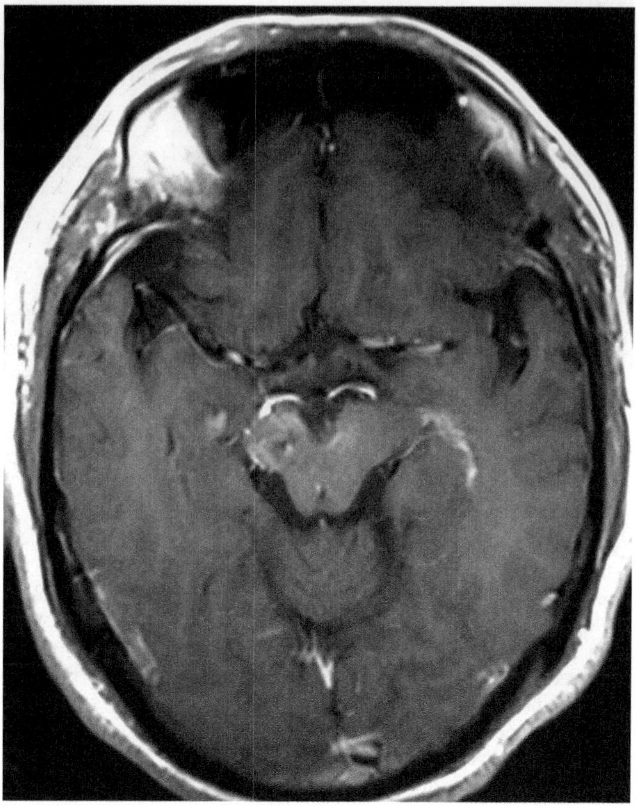

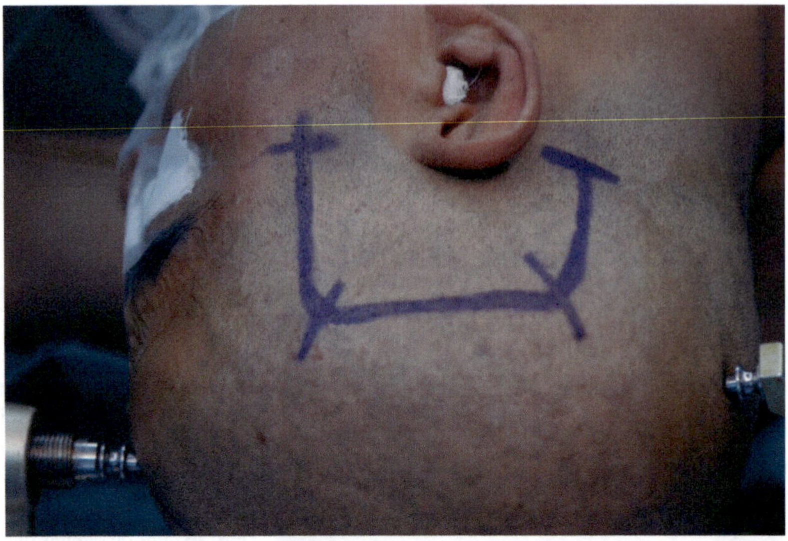

Fig. 27 The patient was placed in a lateral decubitus position. The right "horseshoe" temporal incision was marked

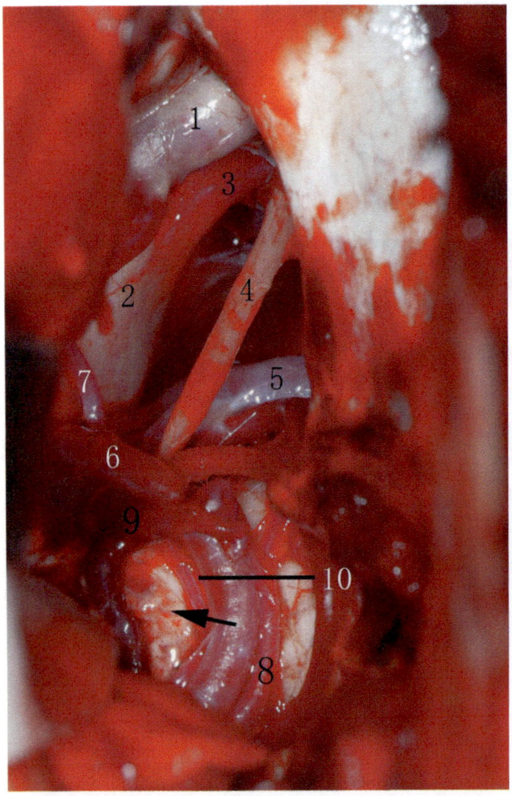

Fig. 28 The hemosiderin tissue (arrow) was visualized following elevation of basal temporal lobe. (*1*) posterior communicating segment of internal carotid artery, (*2*) Median eminence of hypothalamus, (*3*) pituitary stalk, (*4*) 3rd nerve, (*5*) basilar artery, (*6*) posterior communicating segment of posterior cerebral artery, (*7*) posterior communicating artery, (*8*) superior cerebellar artery, (*9*) Basal vein of Rosenthal, and (*10*) medial posterior choroidal artery

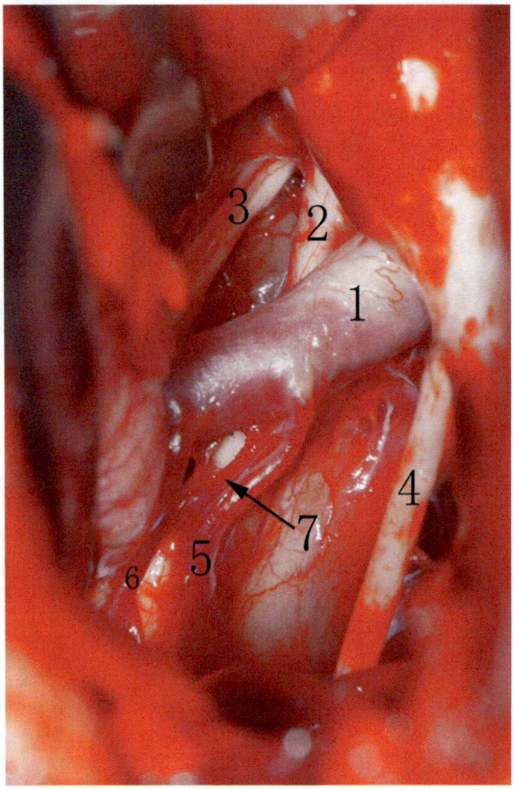

Fig. 29 After malformation removal, peritumoral structures were seen. (*1*) posterior communicating segment of internal carotid artery, (*2*) 2nd nerve, (*3*) olfactory tract, (*4*) 3rd nerve, (*5*) posterior communicating artery, (*6*) anterior choroidal artery, (*7*) premammilary artery

Case 3.3 Giant Ventricular Meningioma

Clinical presentation:

A 16-year-old girl presented with a 2-month history of trembled right hand.

Neurological exam:

Papilledema

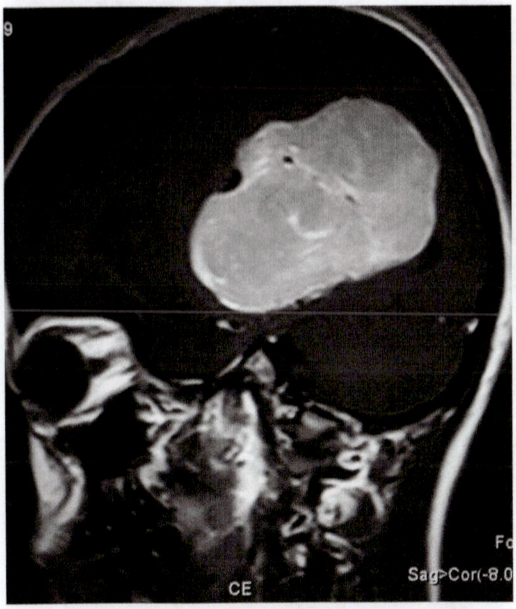

Fig. 30 Pre-op sagittal T1 contrast-enhanced MRI

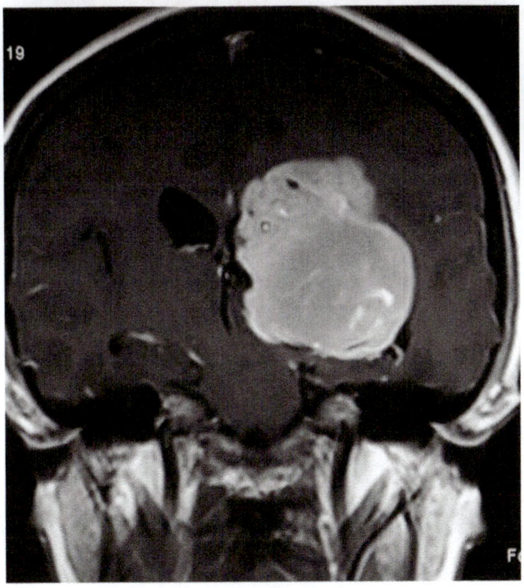

Fig. 31 Pre-op coronal T1 contrast-enhanced MRI

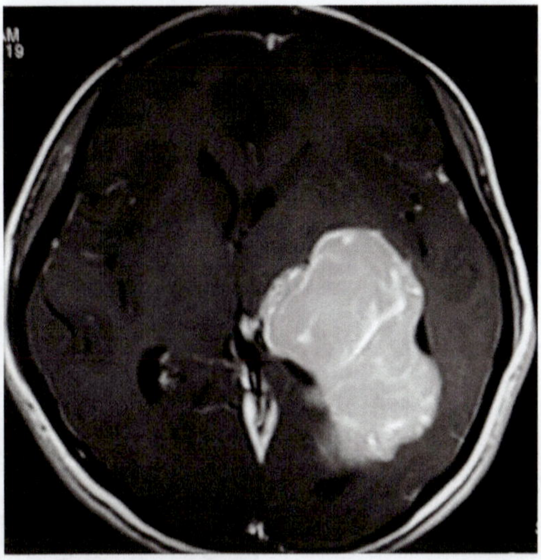

Fig. 32 Pre-op axial T1 contrast-enhanced MRI. Pre-op T1 contrast-enhanced MRI showed a giant solid well-defined intraventricular mass with homogeneous enhancement

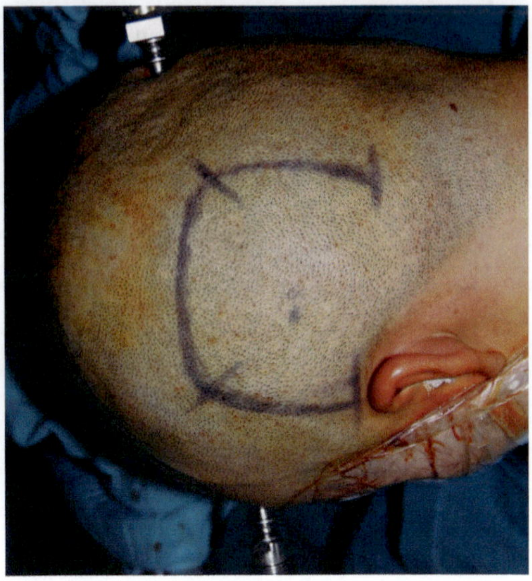

Fig. 33 She was placed in "park-bench" position and "horseshoe" incision was planned

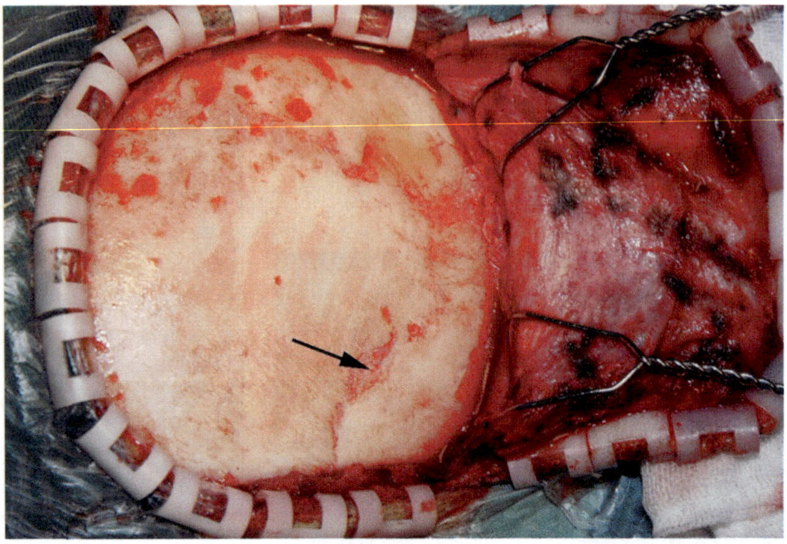

Fig. 34 Skin flap was reflected inferiorly to expose posterior part of squamous suture (arrow) and lambdoid suture

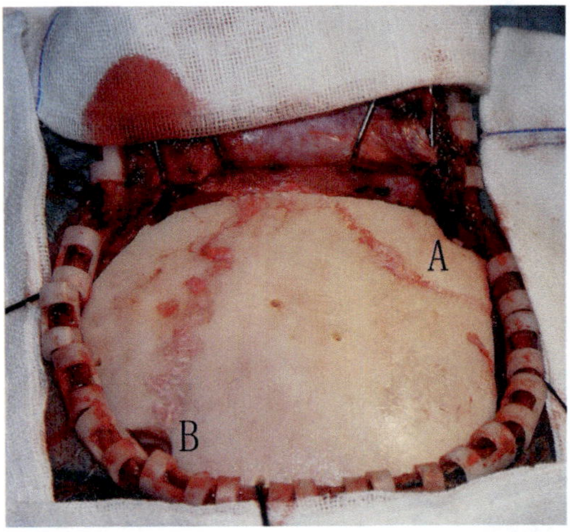

Fig. 35 Three burr holes were made for the craniotomy. (*A*) squamous suture and (*B*) lambdoid suture

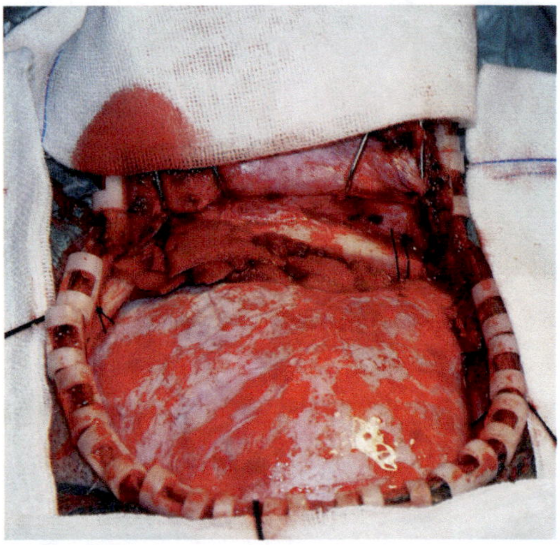

Fig. 36 Tack-up sutures were performed to prevent epidural hematoma

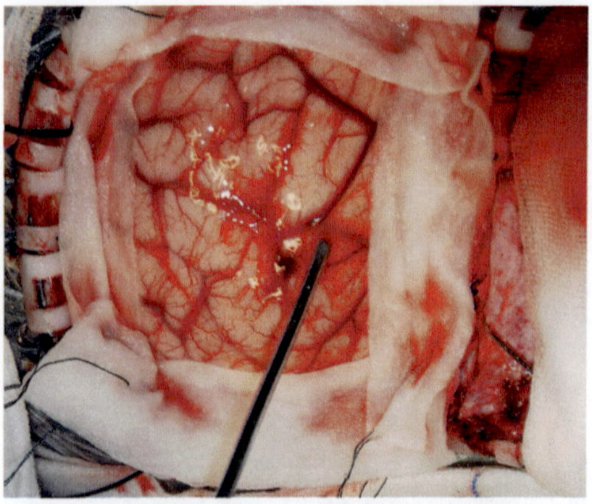

Fig. 37 Dura was opened in stellate fashion

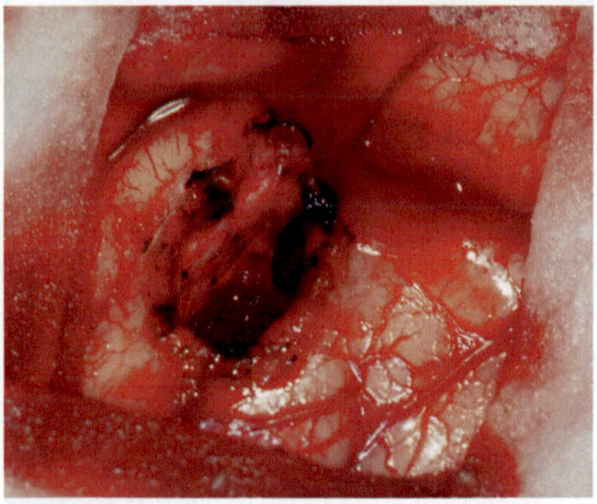

Fig. 38 The trigone of lateral ventricle was approached in transcortical manner that initialized from posterior portion of middle temporal gyrus

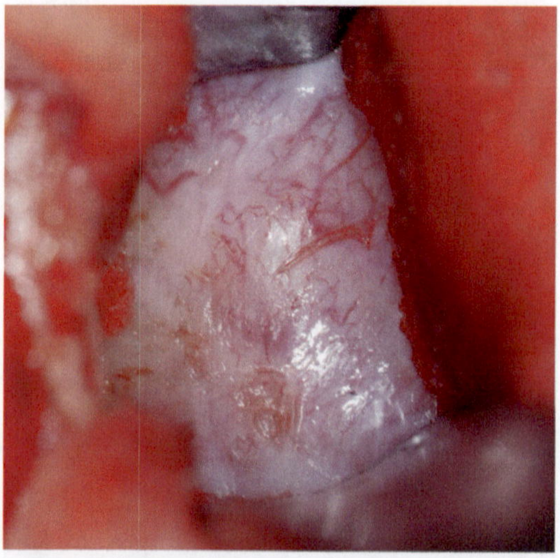

Fig. 39 Tumor was exposed and then resected in a piecemeal fashion

Case 3.4 3rd Ventricle Craniopharyngioma

Clinical presentation:

A 11-month child presented with a 6-month history of headache.

Neurological exam:

Optic atrophy

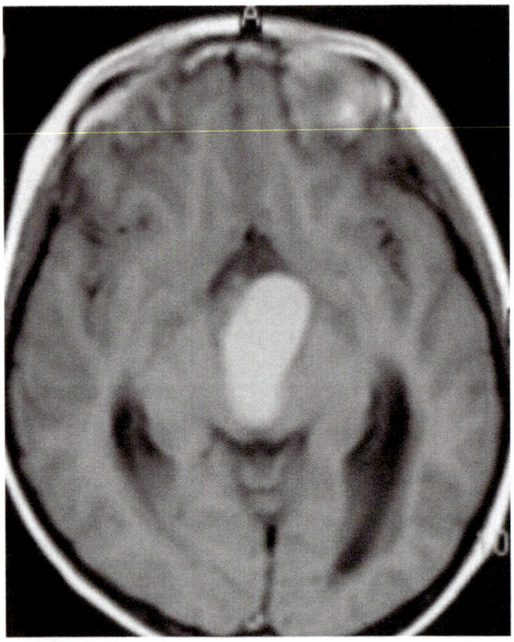

Fig. 40 Pre-op axial T1-weighted MRI

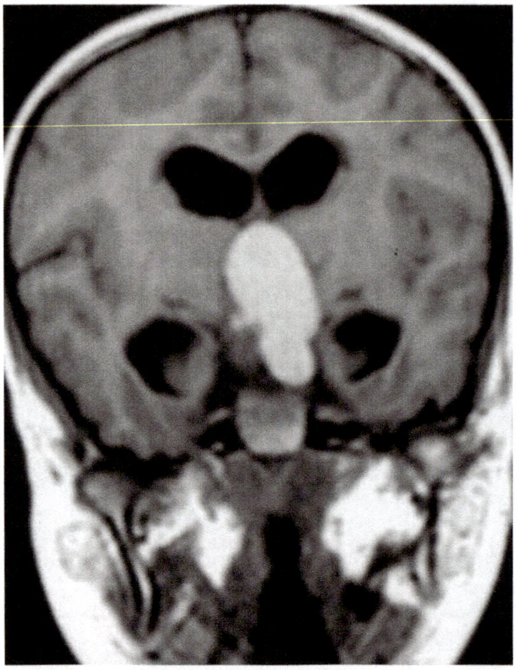

Fig. 41 Pre-op coronal T1-weighted MRI. Pre-op T1-weighted MRI revealed a cystic, well-circumscribed mass eroding suprasellar region and 3rd ventricle

Fig. 42 The patient was placed in supine and incision of frontal craniotomy was labeled

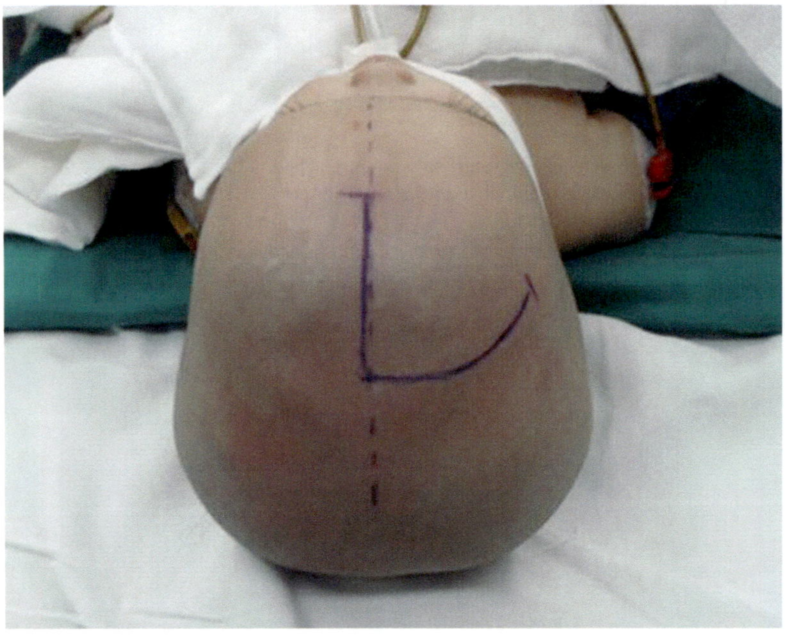

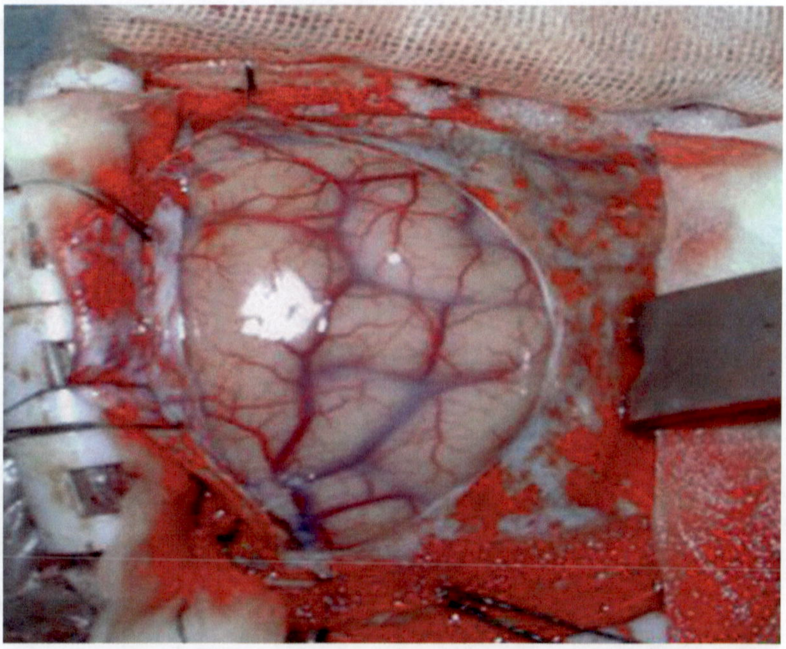

Fig. 43 "C" shape dura opening was achieved with its base at superior sagittal sinus

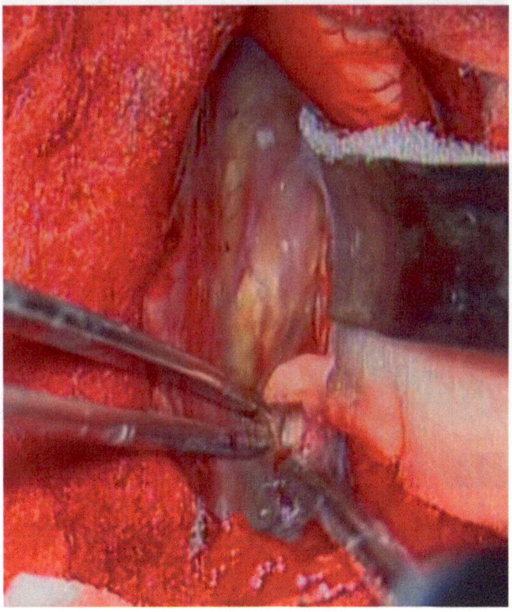

Fig. 44 Interhemispheric fissure was dissected and then the medial frontal lobe was retracted laterally to unveil corpus callosum and bilateral anterior cerebral artery

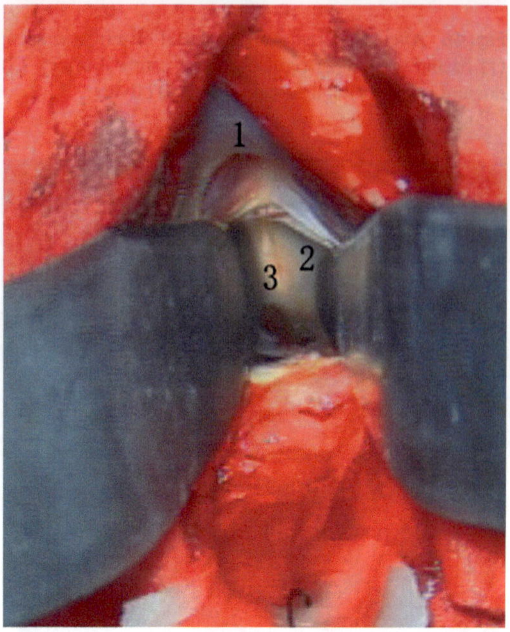

Fig. 45 Callosotomy was performed 2.5 cm posterior to the genu of corpus callosum and incision length was measured at 1 cm. The tenia fornicis was used to open the roof of 3rd ventricle. (*1*) body of corpus callosum, (*2*) septum pellucidum, and (*3*) roof of 3rd ventricle

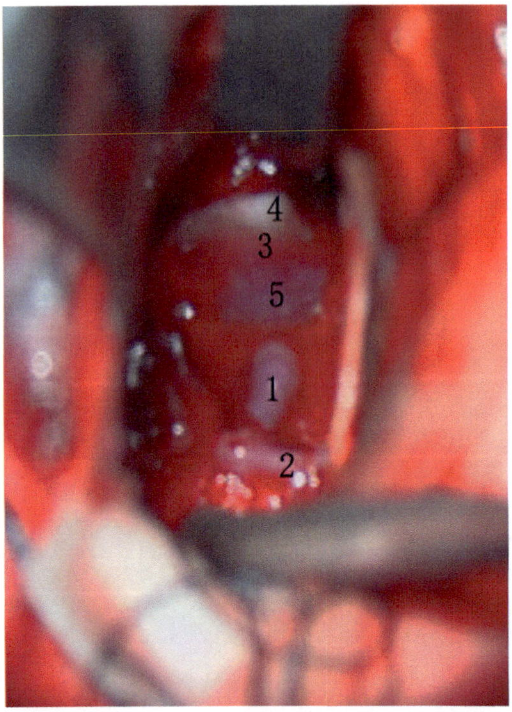

Fig. 46 After mass removal, the basal cistern was seen. (*1*) basilar artery, (*2*) right posterior cerebral artery, (*3*) dorsum sellae, (*4*) optic chiasm, and (*5*) clivus

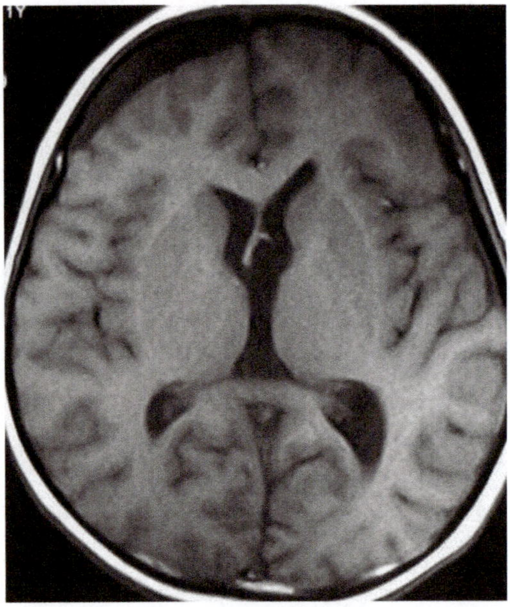

Fig. 47 Post-op axial T1-weighted MRI

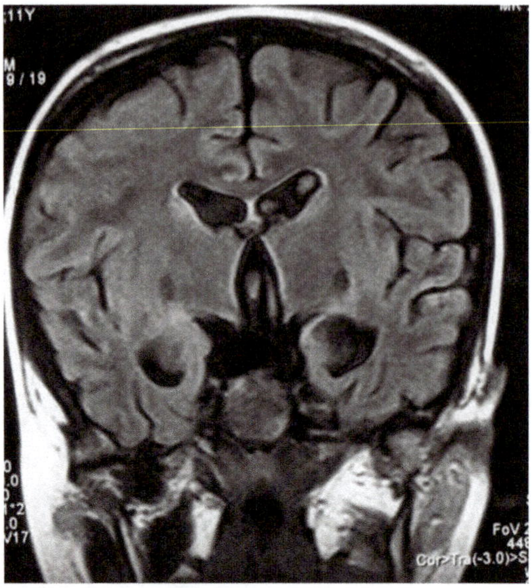

Fig. 48 Post-op coronal T1-weighted MRI

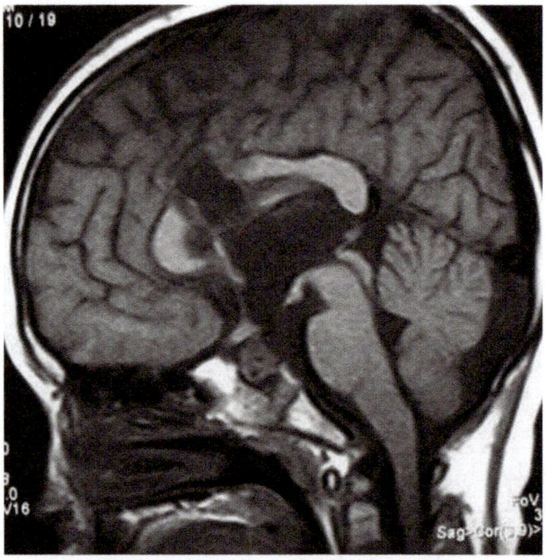

Fig. 49 Post-op sagittal T1-weighted MRI. Post-op T1-weighted MRI revealed total mass resection

Anatomy of Skullbase Arteries, Veins and Surgical Techniques

Xiang'En Shi, Long Wang, and Hai Qian

1 Anatomy of Anterior Circulation Arteries

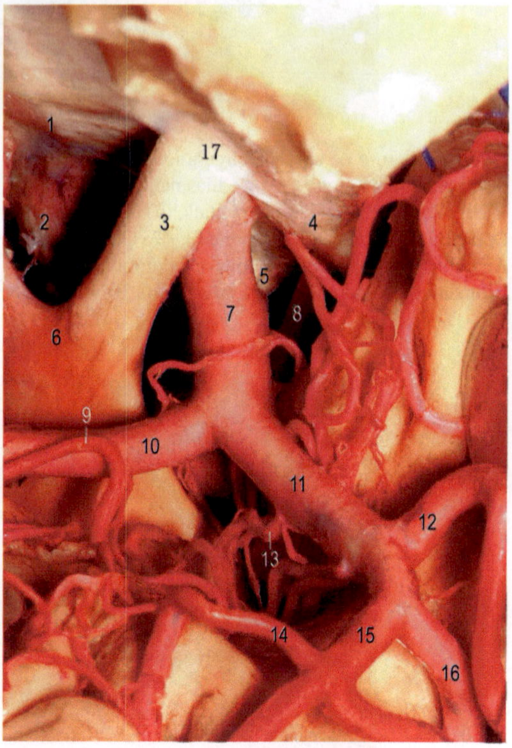

Fig. 1 Overview of anterior circulation arteries. (*1*) Tuberculum sellae, (*2*) Pituitary stalk, (*3*) Optic nerve, (*4*) Anterior clinoid process, (*5*) Posterior clinoid process, (*6*) Optic chiasm, (*7*) Internal carotid artery, (*8*) Oculomotor nerve, (*9*) Recurrent artery of Heubner, (*10*) Anterior cerebral artery, (*11*) Middle cerebral artery, (*12*) Anterior temporal artery, (*13*) Lateral lenticulostriate artery, (*14*) Lateral orbitofrontal artery, (*15*) Frontal trunk, (*16*) Temporal trunk, (*17*) Falciform ligament

X. Shi (✉) · L. Wang · H. Qian
Neurological Surgery, SanBo Brain Hospital, Capital Medical University, Beijing, China
e-mail: shixen@ccmu.edu.cn; wanglong@ccmu.edu.cn; qianh@ccmu.edu.cn

X. Shi et al. (eds.), *Microneurosurgical Anatomy and Surgical Technique*,
https://doi.org/10.1007/978-981-19-8273-6_4

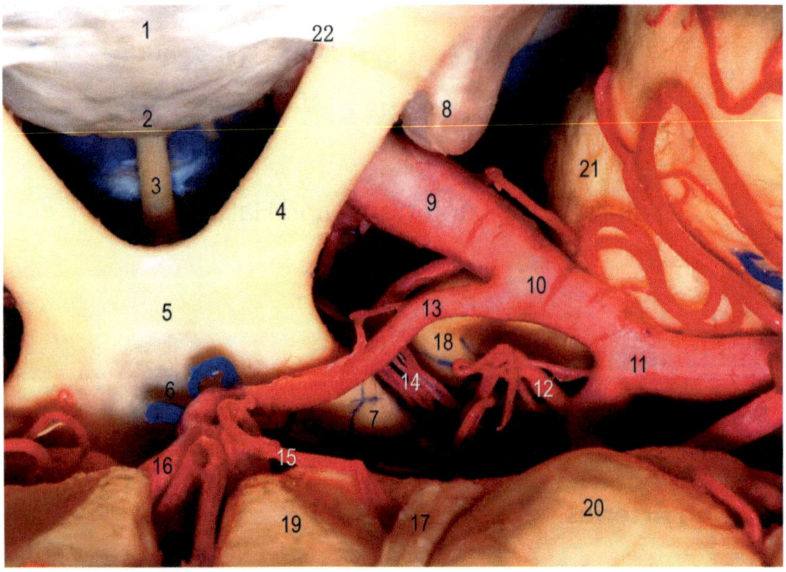

Fig. 2 Exposure of anterior circulation arteries through subfrontal approach. (*1*) Chiasmatic sulcus, (*2*) Tuberculum Sellae, (*3*) Pituitary stalk, (*4*) Optic nerve, (*5*) Optic chiasm, (*6*) Lamina terminalis, (*7*) Optic tract, (*8*) Anterior clinoid process, (*9*) Internal carotid artery, (*10*) Bifurcation of internal carotid artery, (*11*) Bifurcation of middle cerebral artery, (*12*) Lateral lenticulostriate artery, (*13*) Anterior cerebral artery, (*14*) Anterior choroidal artery, (*15*) Medial orbitofrontal artery, (*16*) Anterior communicating artery, (*17*) Olfactory tract, (*18*) Uncus, (*19*) Gyrus rectus, (*20*) Posterior orbital gyrus, (*21*) Parahippocampal gyrus, (*22*) Falciform ligament

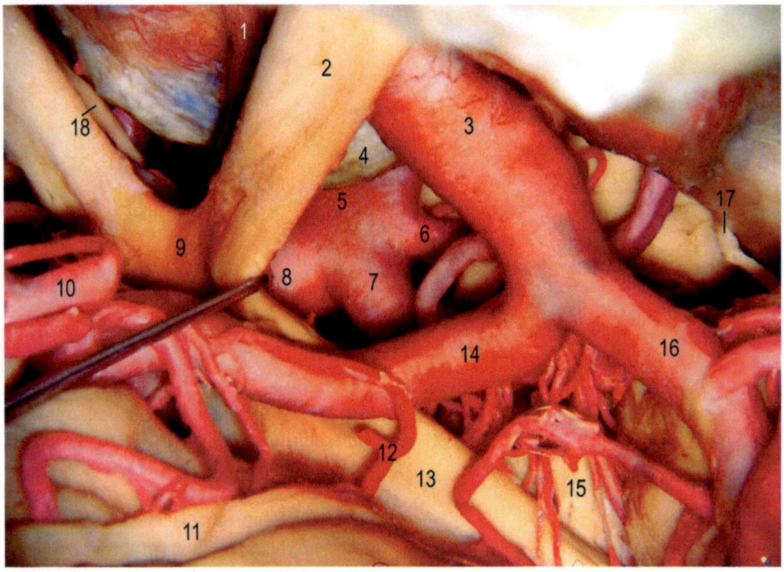

Fig. 3 Exposure of basilar apex complex through subfrontal approach. (*1*) Pituitary stalk, (*2*) Optic nerve, (*3*) Internal carotid artery, (*4*) Posterior clinoid process, (*5*) Basilar artery, (*6*) Right superior cerebellar artery, (*7*) Right posterior cerebral artery, (*8*) Left posterior cerebral artery, (*9*) Optic chiasm, (*10*) Left postcommunicating segment of anterior cerebral artery, (*11*) Olfactory tract, (*12*) Medial orbitofrontal artery, (*13*) Optic tract, (*14*) Precommunicating segment of anterior cerebral artery, (*15*) Uncus, (*16*) Middle cerebral artery, (*17*) Trochlear nerve, (*18*) Left oculomotor nerve

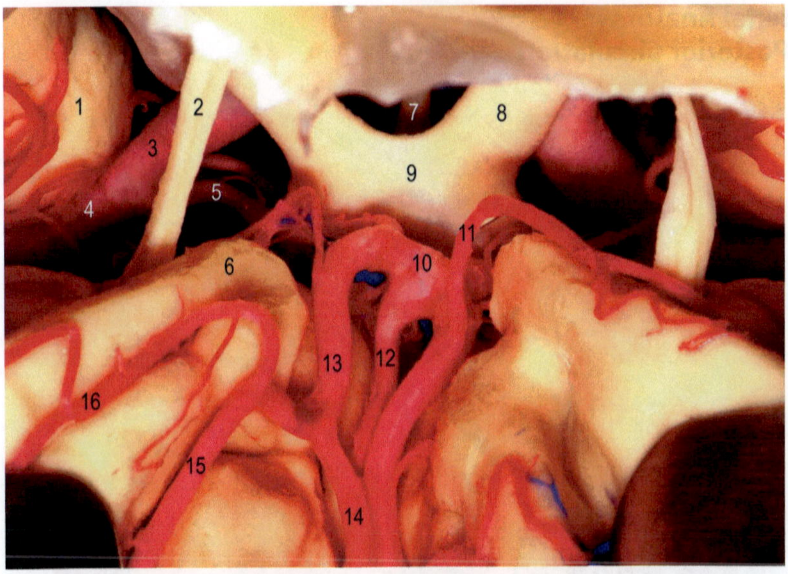

Fig. 4 Exposure of anterior circulation arteries through frontal interhemispheric approach. (*1*) Temporal pole, (*2*) Olfactory tract, (*3*) Internal carotid artery, (*4*) Middle cerebral artery, (*5*) Anterior cerebral artery, (*6*) Gyrus rectus, (*7*) Pituitary stalk, (*8*) Optic nerve, (*9*) Optic chiasm, (*10*) Anterior communicating artery, (*11*) Medial orbitofrontal artery, (*12*) Median artery of the corpus callosum/accessory anterior cerebral artery, (*13*) Postcommunicating segment of anterior cerebral artery, (*14*) Pericallosal artery, (*15*) Callosomarginal artery, (*16*) Anterior medial frontal artery

Fig. 5 Inferior view of anterior circulation arteries. (*1*) Olfactory tract, (*2*) Optic nerve, (*3*) Right anterior cerebral artery, (*4*) Middle cerebral artery, (*5*) Internal carotid artery, (*6*) Posterior communicating artery, (*7*) Optic tract, (*8*) Optic chiasm, (*9*) Infundibular stalk, (*10*) Tuber cinereum, (*11*) Middle (Paramedian) perforated substance, (*12*) Precommunicating segment of posterior cerebral artery, (*13*) Mamillary body, (*14*) Uncus, (*15*) Postcommunicating segment of posterior cerebral artery, (*16*) Cerebral peduncle, (*17*) Parahippocampal gyrus, (*18*) Pons, (*19*) Premamillary artery

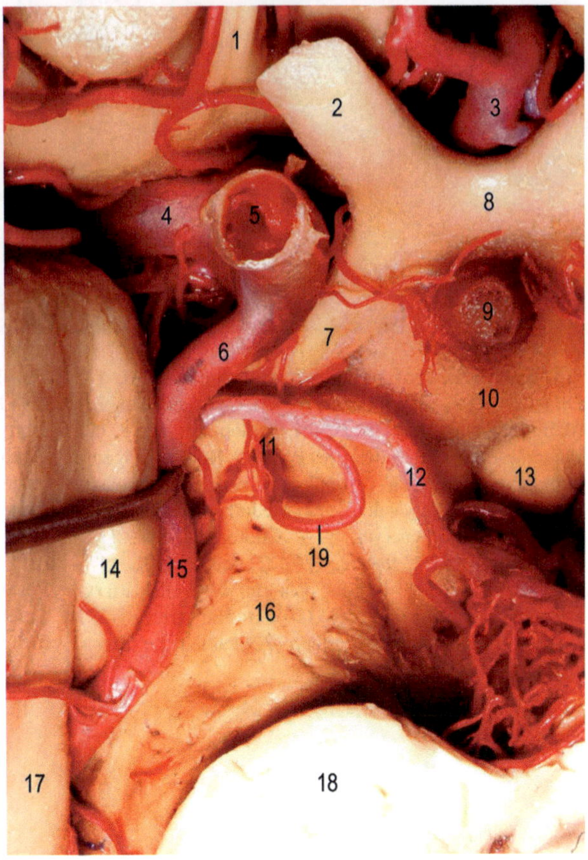

2 Arterial Anatomy of Posterior Circulation

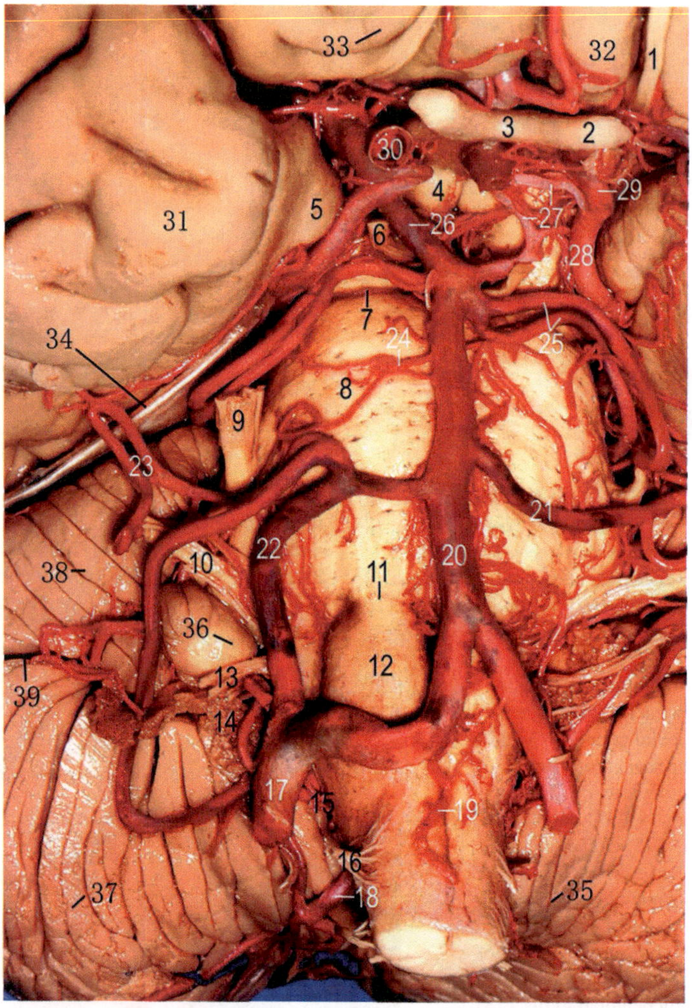

Fig. 6 Overview of posterior circulation arteries. (*1*) Olfactory tract, (*2*) Optic nerve, (*3*) Optic chiasm, (*4*) Mamillary body, (*5*) Uncus, (*6*) Oculomotor nerve, (*7*) Pontomesencephalic sulcus, (*8*) Pons, (*9*) Roots of trigeminal nerve, (*10*) Facial–vestibulocochlear nerve complex, (*11*) Pontomedullary sulcus, (*12*) Medulla oblongata, (*13*) Glossopharyngeal nerves, (*14*) Choroid plexus, (*15*) Cerebellomedullary fissure, (*16*) C1 nerve root, (*17*) Vertebral artery, (*18*) Inferior posterior cerebellar artery, (*19*) Anterior spinal artery, (*20*) Basilar artery, (*21*) Inferior anterior cerebellar artery, (*22*) Parabasilar artery, (*23*) Rostral trunk of inferior anterior cerebellar artery, (*24*) Pontine arteries, (*25*) Duplicated superior cerebellar artery, (*26*) Precommunicating segment of right posterior cerebral artery, (*27*) Precommunicating segment of left posterior cerebral artery, (*28*) Postcommunicating segment of left posterior cerebral artery, (*29*) Left posterior communicating artery, (*30*) Right internal carotid artery, (*31*) Parahippocampal gyrus, (*32*) Gyrus rectus, (*33*) Posterior orbital gyrus, (*34*) Tentorial incisure, (*35*) Cerebellar tonsil, (*36*) Flocullus, (*37*) Biventral lobule, (*38*) Sample lobule, (*39*) Horizontal fissure

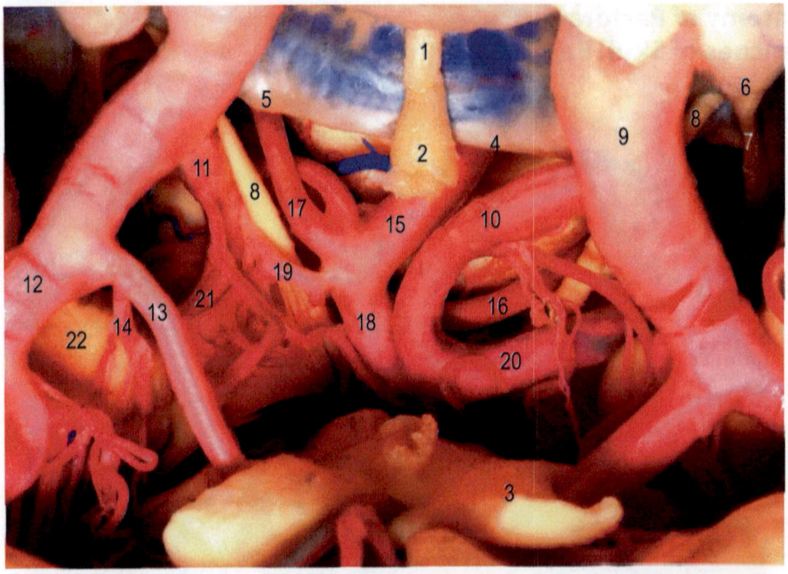

Fig. 7 Exposure of posterior circulation arteries through subfrontal approach. (*1*) Pituitary stalk, (*2*) Medial eminence, (*3*) Optic nerve, (*4*) Dorsum Sellae, (*5*) Posterior clinoid process, (*6*) Anterior clinoid process, (*7*) Anterior petroclinoid fold, (*8*) Oculomotor nerve, (*9*) Internal carotid artery, (*10*) Right posterior communicating artery, (*11*) Left posterior communicating artery, (*12*) Middle cerebral artery, (*13*) Anterior cerebral artery, (*14*) Anterior choroidal artery, (*15*) Basilar trunk, (*16*) Right superior cerebellar artery, (*17*) Left superior cerebellar artery, (*18*) Precommunicating segment of right posterior cerebral artery, (*19*) Precommunicating segment of left posterior cerebral artery, (*20*) Postcommunicating segment of right posterior cerebral artery, (*21*) Postcommunicating segment of left posterior cerebral artery, (*22*) Uncus

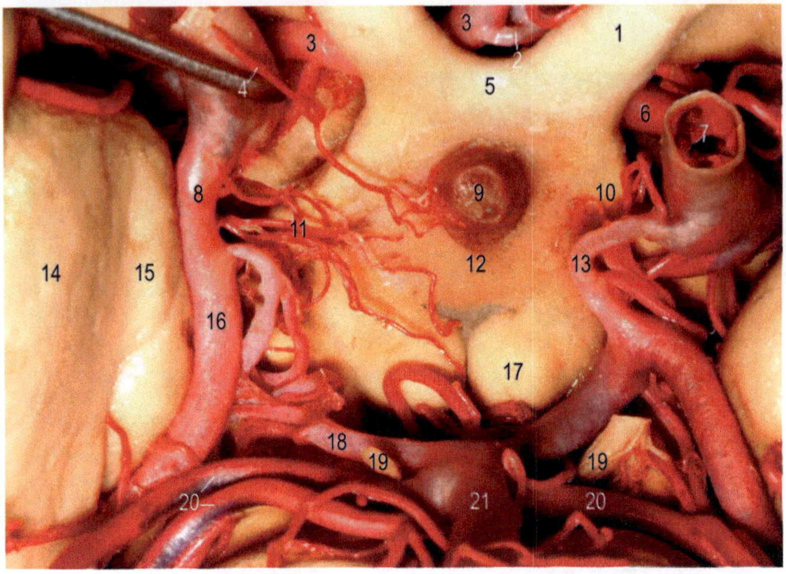

Fig. 8 Inferior view of posterior circulation arteries. (*1*) Optic nerve, (*2*) Anterior communicating artery, (*3*) Right anterior cerebral artery, (*4*) Superior hypophysial artery, (*5*) Optic chiasm, (*6*) Left anterior cerebral artery, (*7*) Internal carotid artery, (*8*) Right posterior communicating artery, (*9*) Infundibular stalk, (*10*) Optic tract, (*11*) Premamillary artery, (*12*) Tuber cinereum, (*13*) Left posterior communicating artery, (*14*) Parahippocampal gyrus, (*15*) Uncus, (*16*) Postcommunicating segment of posterior cerebral artery, (*17*) Mamillary body, (*18*) Precommunicating segment of posterior cerebral artery, (*19*) Oculomotor nerve, (*20*) Superior cerebellar artery, (*21*) Basilar artery

3 Anatomy of Perforating Branches of Skull Base Arteries

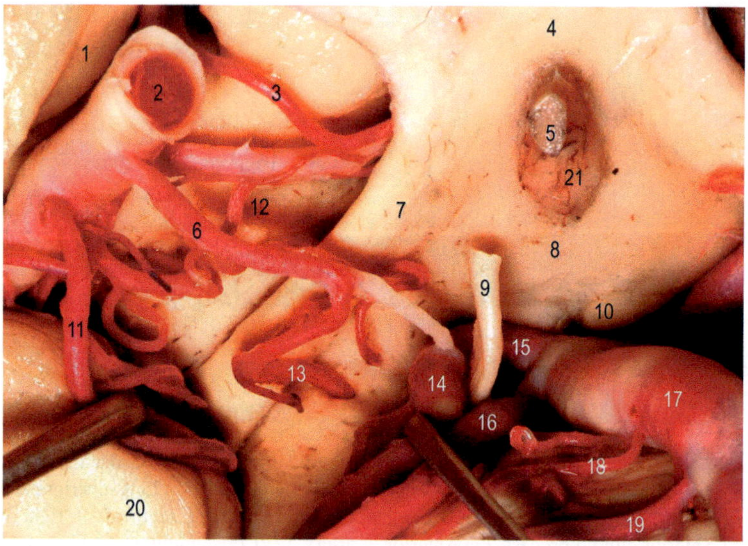

Fig. 9 Overview of perforating branches from skull base arteries. (*1*) Posteromedial orbital lobule, (*2*) Internal carotid artery, (*3*) Medial orbitofrontal artery, (*4*) Optic chiasm, (*5*) Pituitary stalk, (*6*) Posterior communicating artery, (*7*) Optic tract, (*8*) Tuber cinereum, (*9*) Oculomotor nerve, (*10*) Mamillary body, (*11*) Anterior choroidal artery, (*12*) Anterior perforated substance, (*13*) Premamillary artery, (*14*) Postcommunicating segment of posterior cerebral artery, (*15*) Precommunicating segment of posterior cerebral artery, (*16*) Superior cerebellar artery, (*17*) Basilar artery, (*18*) Pontine arteries, (*19*) Anterior inferior cerebellar artery, (*20*) Uncus, (*21*) Median eminence

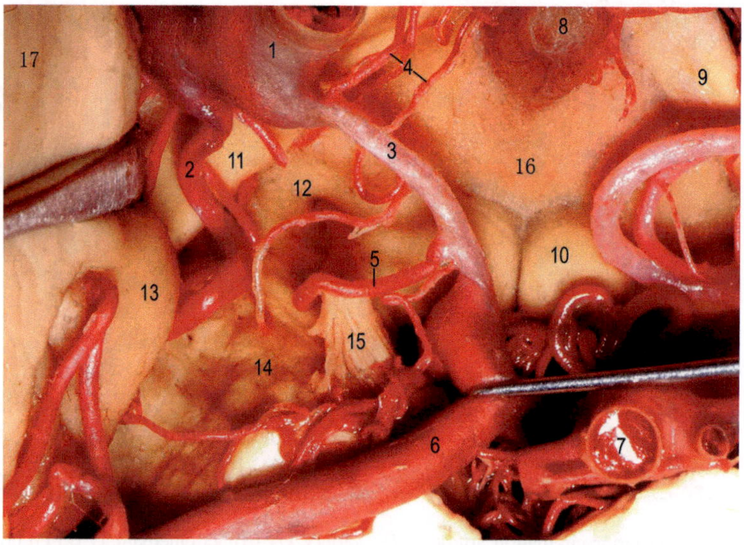

Fig. 10 Inferior view of perforating branches from posterior communicating artery. (*1*) Internal carotid artery, (*2*) Anterior choroidal artery, (*3*) Posterior communicating artery, (*4*) Tuberoinfundibular artery, (*5*) Premamillary artery, (*6*) Postcommunicating segment of posterior cerebral artery, (*7*) Basilar artery, (*8*) Infundibular stalk, (*9*) Left optic tract, (*10*) Mammillary body, (*11*) Right optic tract, (*12*) Middle (Paramedian) perforated substance, (*13*) Uncus, (*14*) Cerebral peduncle, (*15*) Oculomotor nerve, (*16*) Tuber cinereum, (*17*) Parahippocampal gyrus

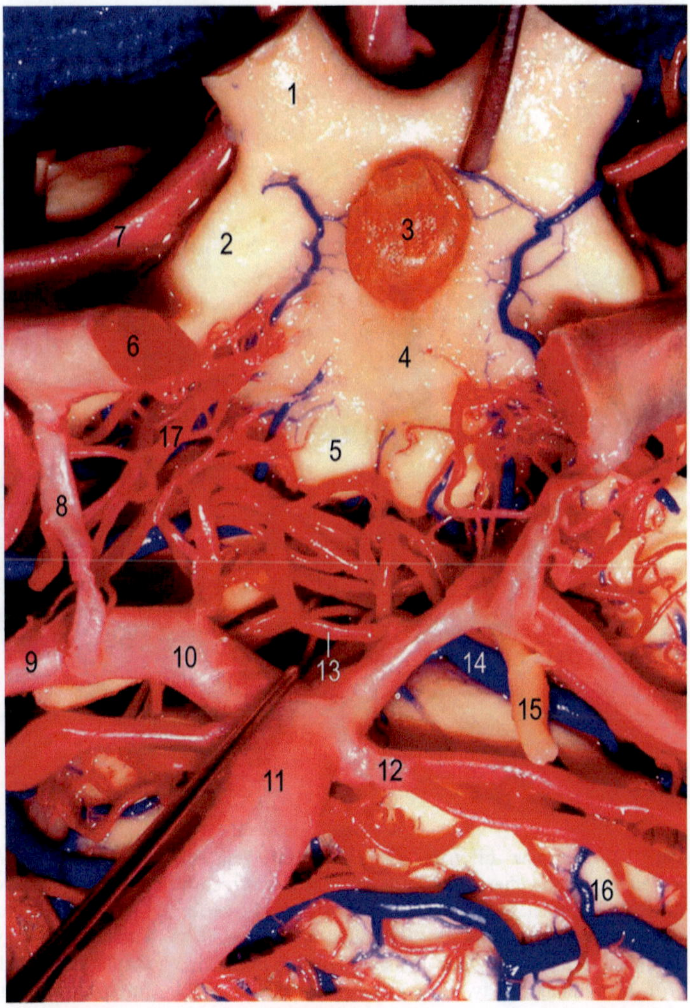

Fig. 11 Inferior view of perforating branches from proximal posterior cerebral artery. (*1*) Optic nerve, (*2*) Optic tract, (*3*) Infundibular stalk, (*4*) Tuber cinereum, (*5*) Mammillary body, (*6*) Internal carotid artery, (*7*) Anterior cerebral artery, (*8*) Posterior communicating artery, (*9*) Postcommunicating segment of posterior cerebral artery, (*10*) Precommunicating segment of posterior cerebral artery, (*11*) Basilar artery, (*12*) Superior cerebellar artery, (*13*) Posterior thalamoperforating arteries, (*14*) Basal vein of Rosenthal, (*15*) Oculomotor nerve, (*16*) Pons, (*17*) Premamillary arteries

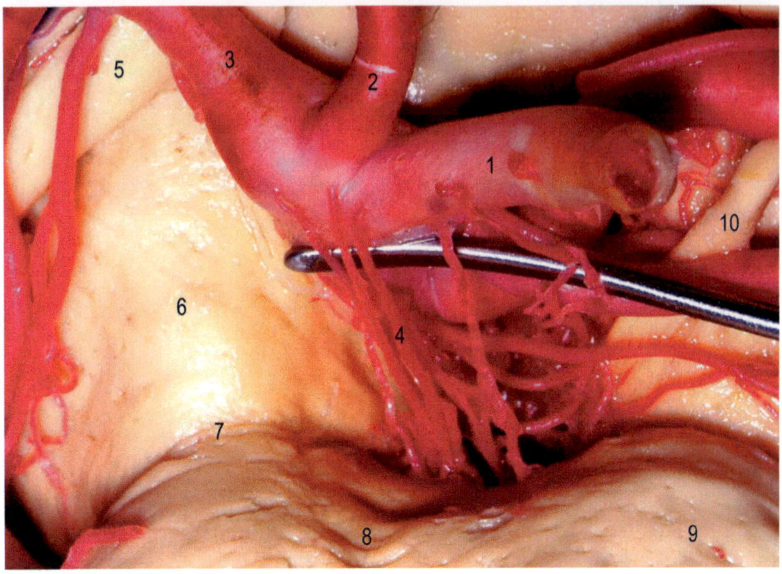

Fig. 12 Perforating branches from upper basilar artery. (*1*) Basilar artery, (*2*) Superior cerebellar artery, (*3*) Posterior cerebral artery, (*4*) Posterior thalamoperforating arteries, (*5*) Optic tract, (*6*) Cerebral peduncle, (*7*) Pontomesencephalic sulcus, (*8*) Basilar sulcus, (*9*) Pons, (*10*) Oculomotor nerve

4 Anatomy of Anterior Venous System

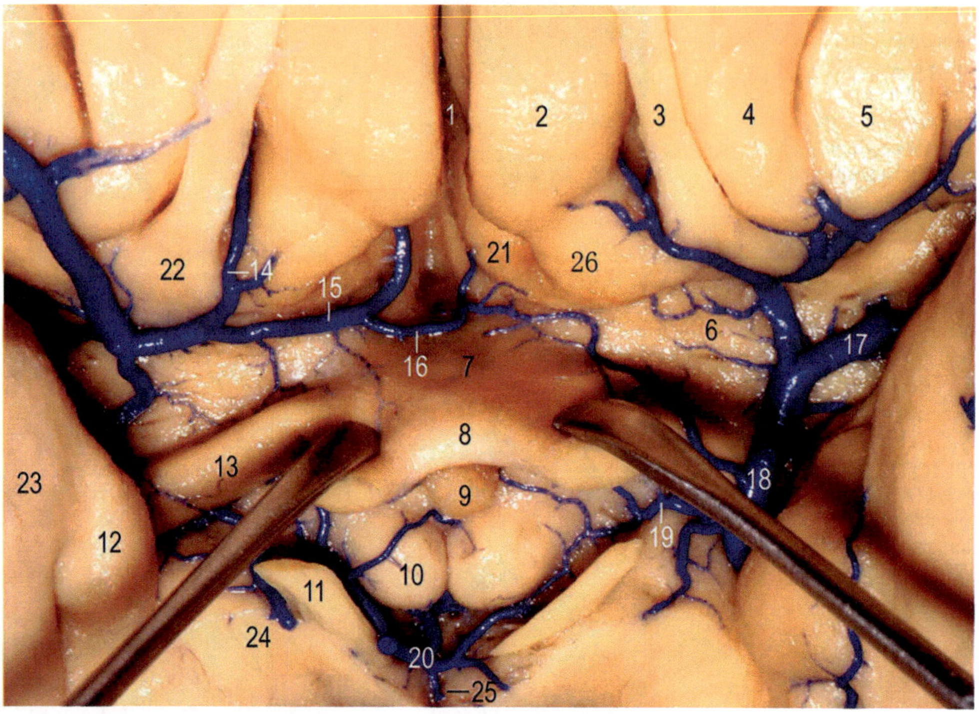

Fig. 13 Overview of tributaries of basal vein of Rosenthal. (*1*) Interhemispheric fissure, (*2*) Gyrus rectus, (*3*) Olfactory tract, (*4*) Medial orbital gyrus, (*5*) Posterior orbital gyrus, (*6*) Anterior perforated substance, (*7*) Lamina terminalis, (*8*) Optic chiasm, (*9*) Tuber cinereum, (*10*) Mammillary body, (*11*) Oculomotor nerve, (*12*) Uncus, (*13*) Optic tract, (*14*) Olfactory vein, (*15*) Anterior cerebral vein, (*16*) Anterior communicating vein, (*17*) Deep middle cerebral vein, (*18*) Basal vein of Rosenthal, (*19*) Hypothalamic vein, (*20*) Peduncular vein, (*21*) Paraterminal gyrus, (*22*) Olfactory trigone, (*23*) Parahippocampal gyrus, (*24*) Cerebral peduncle, (*25*) Interpeduncular fossa, (*26*) Paraolfactory gyrus

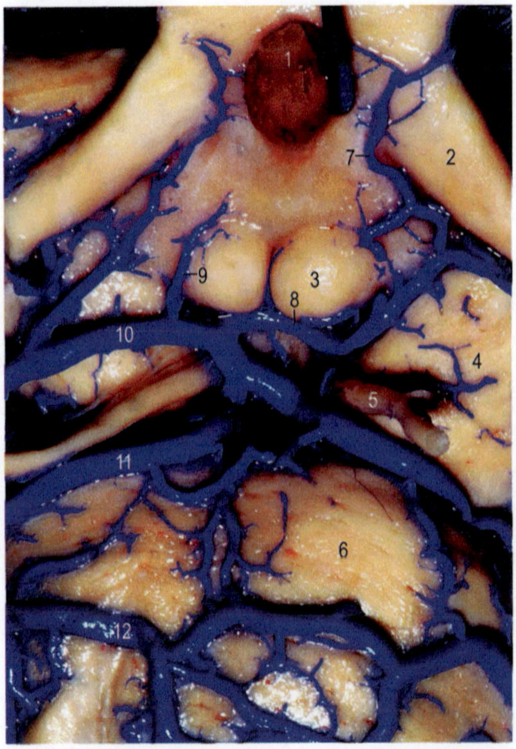

Fig. 14 Inferior view of mesencephalic and pontine veins. (*1*) Infundibular vein Infundibulum, (*2*) Optic tract, (*3*) Mammillary body, (*4*) Cerebral peduncle, (*5*) Oculomotor nerve, (*6*) Pons, (*7*) Striate segment of basal vein of Rosenthal, (*8*) Left peduncular vein, (*9*) Hypothalamic vein, (*10*) Right peduncular vein, (*11*) Vein of pontomesencephalic sulcus, (*12*) Transverse pontine vein

Case 4.1 Internal Carotid-Posterior Communicating Artery Aneurysm

Clinical presentation:

A 65-year-old gentleman presented with a 6-month history of severe headache.

Neurological exam:

Decreased right visual acuity, right nasal hemianopsia

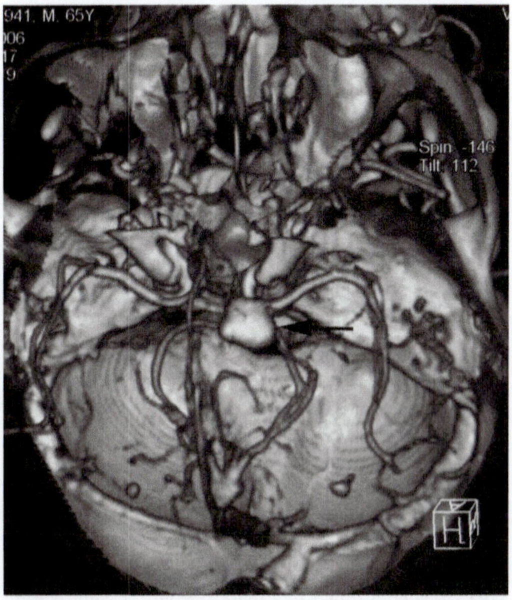

Fig. 15 Pre-op CT angiography revealed a large wide-neck aneurysm of posterior communicating segment of internal carotid artery (arrow)

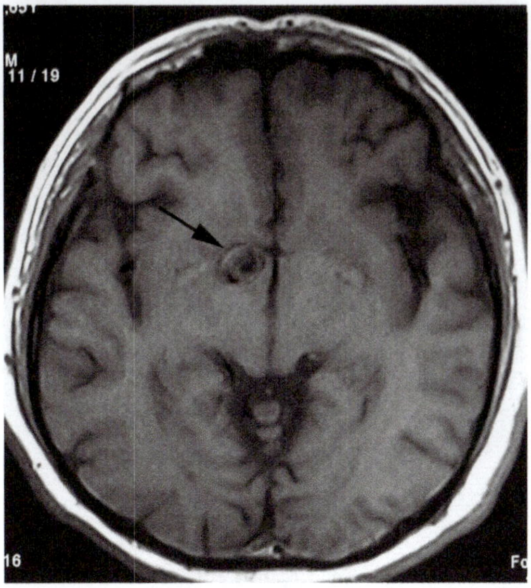

Fig. 16 Pre-op axial T1-weighted MRI

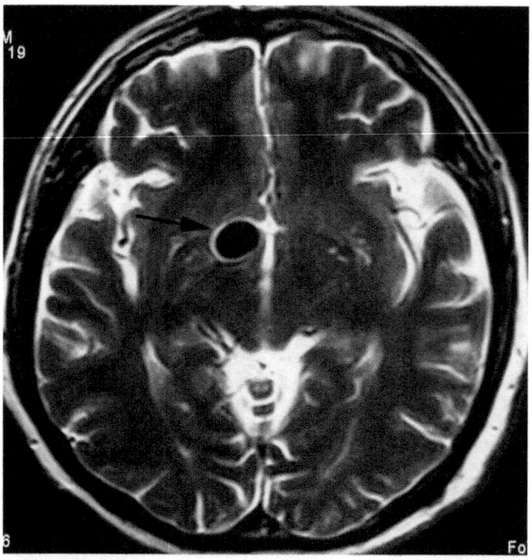

Fig. 17 Pre-op axial T2-weighted MRI. Pre-op axial MRI revealed a round-like mass (arrow) of right carotid cistern with a flow void sign

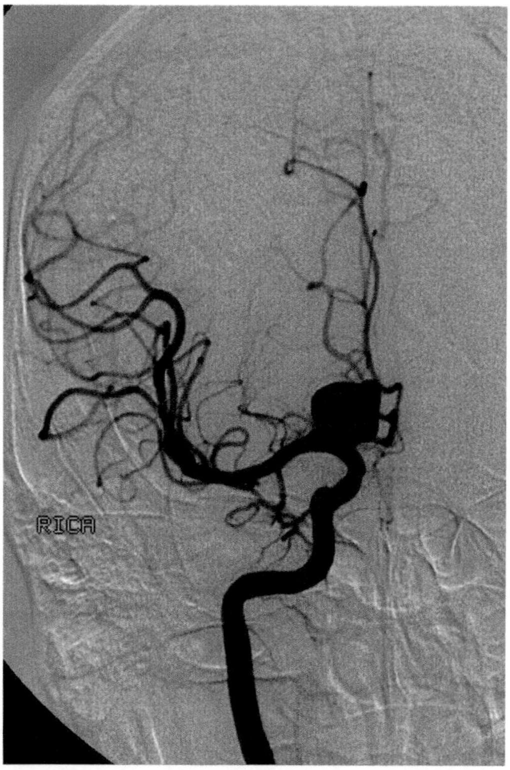

Fig. 18 Pre-op anteroposterior view of right internal carotid artery angiography

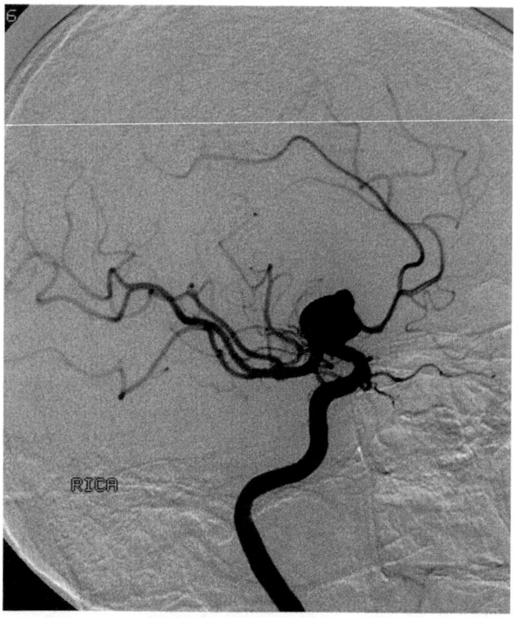

Fig. 19 Pre-op lateral view of right internal carotid artery angiography. Pre-op right internal carotid artery angiography confirmed a vascular abnormality indicating a large C7 aneurysm with formation of a daughter sac at the dome

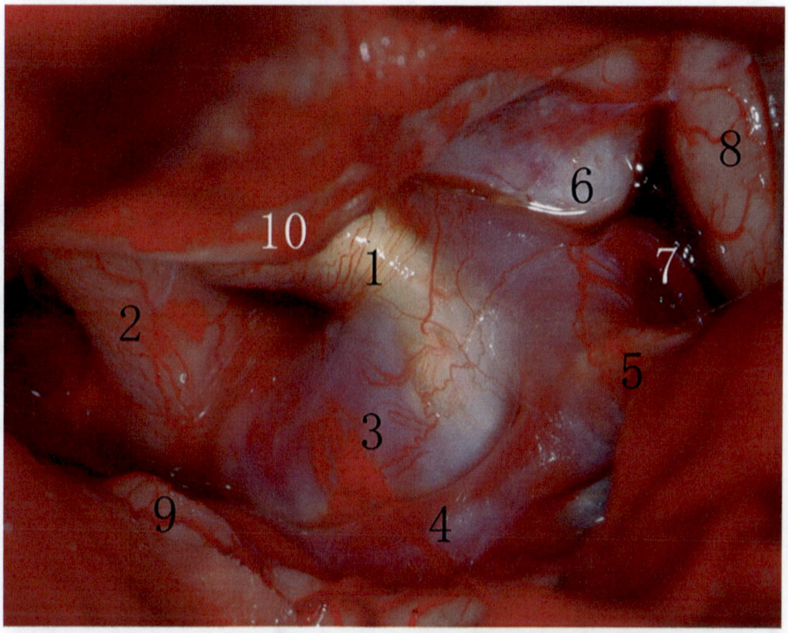

Fig. 20 The aneurysm was fully exposed by pterional approach, dissection of sylvian, carotid, and chiasmatic (suprasellar) cistern. (*1*) Supraclinoid segment of internal carotid artery, (*2*) right optic nerve, (*3*) aneurysm, (*4*) precommunicating segment of anterior cerebral artery, (*5*) sphenoidal segment of middle cerebral aneurysm, (*6*) posterior clinoid process, (*7*) anterior choroidal artery, (*8*) uncus, (*9*) pars orbitalis, and (*10*) anterior clinoid process

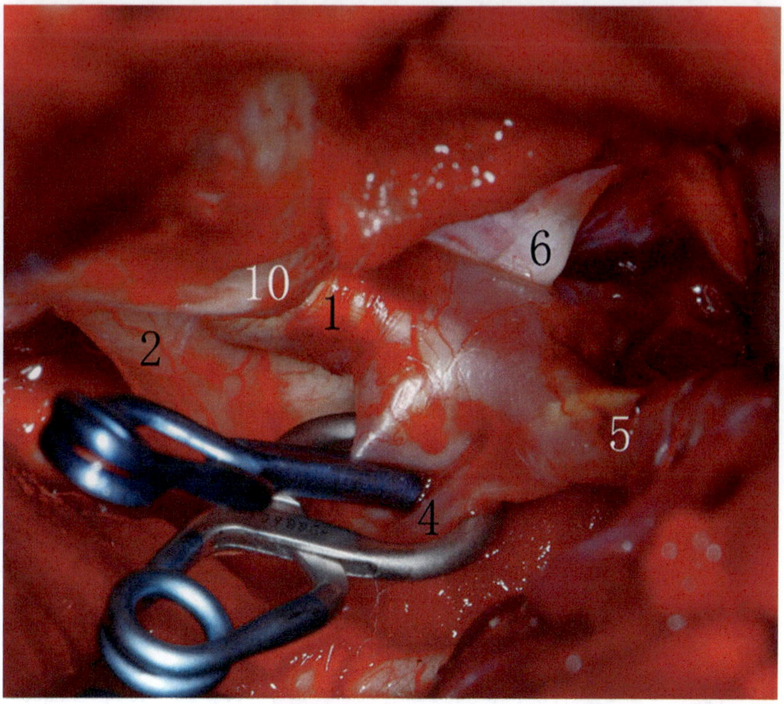

Fig. 21 The aneurysm was repaired by tandem clip technique. (*1*) Supraclinoid segment of internal carotid artery, (*2*) right optic nerve, (*4*) precommunicating segment of anterior cerebral artery,(*5*) sphenoidal segment of middle cerebral aneurysm,(*6*) posterior clinoid process, and (*10*) anterior clinoid process

Case 4.2 Giant Thrombosed Anterior Circulation Aneurysm

Clinical presentation:

A 20-year-old boy presented with a 2-month history of severe headache and visual loss.

Neurological exam:

Decreased bilateral visual acuity, right ptosis

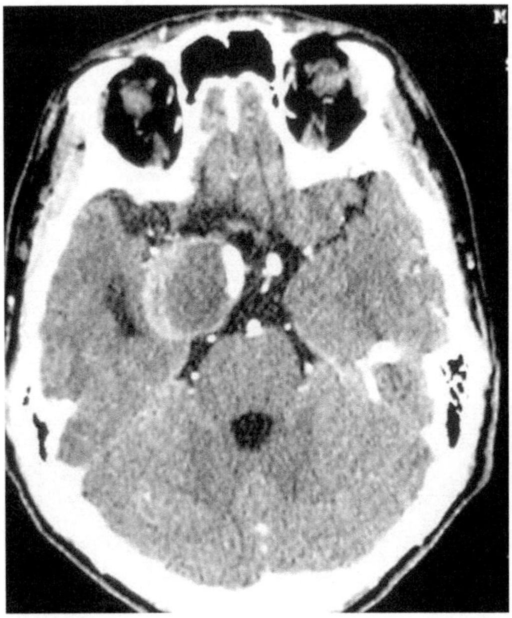

Fig. 22 Pre-op axial CT showed a round mass of right parasellar region

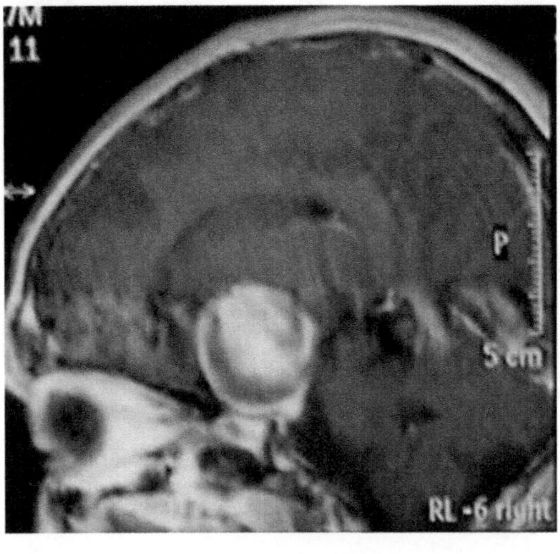

Fig. 23 Pre-op sagittal Gd-enhanced MRI revealed a giant partially thrombosed aneurysm with ring enhancement

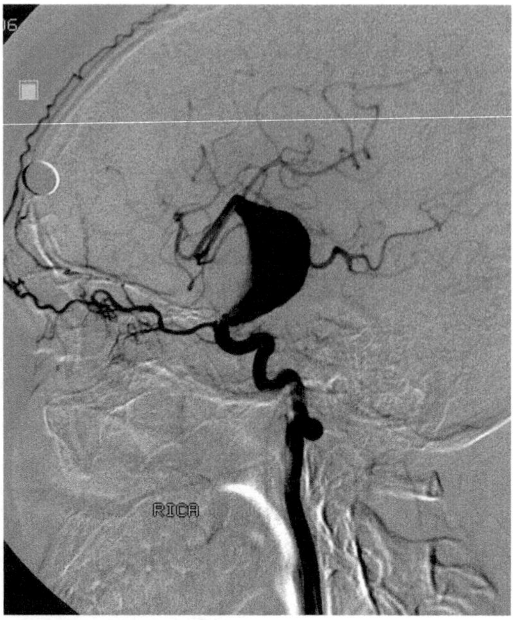

Fig. 24 Pre-op lateral view of right internal carotid artery angiography

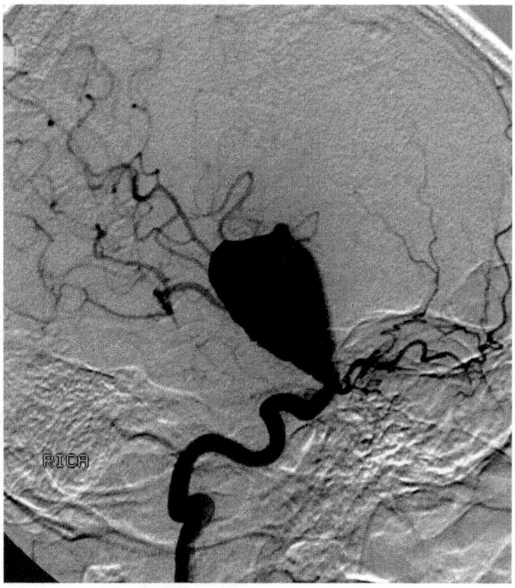

Fig. 25 Pre-op oblique view of right internal carotid artery angiography. Pre-op right internal carotid artery angiography illustrated that the supraclinoidal segment of internal carotid artery, proximal anterior and middle cerebral artery were got involved by this lesion

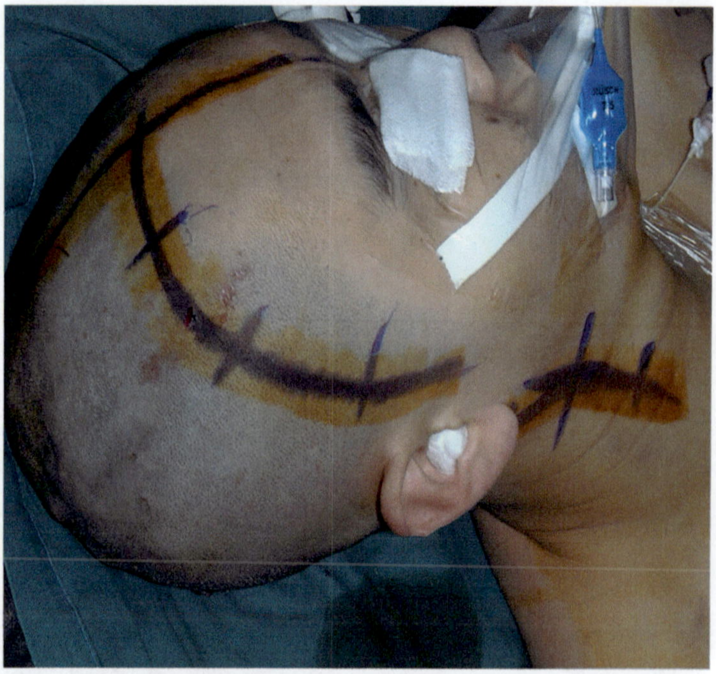

Fig. 26 The right frontotemporal craniotomy and cervical carotid exposure was planned

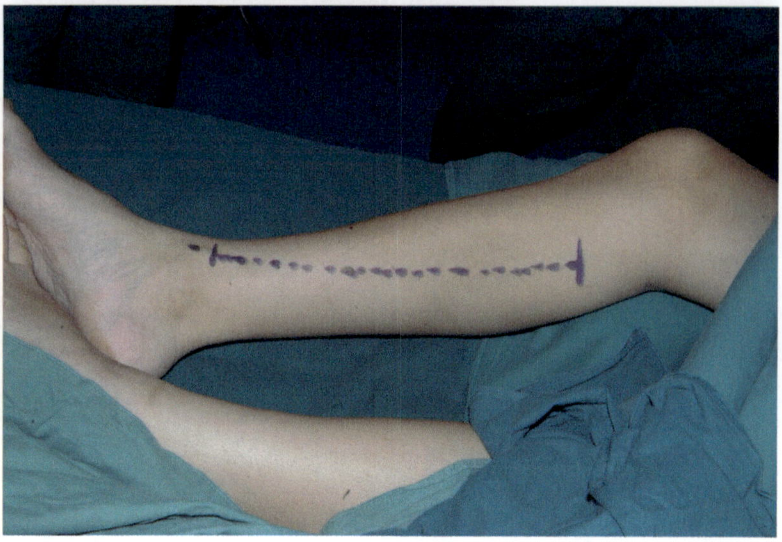

Fig. 27 The course of great saphenous vein was marked

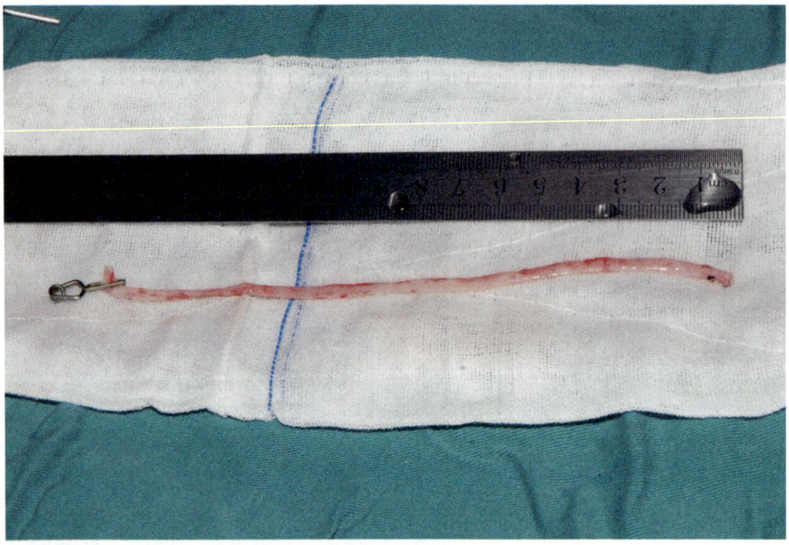

Fig. 28 Venous graft was harvested for 15–20 cm

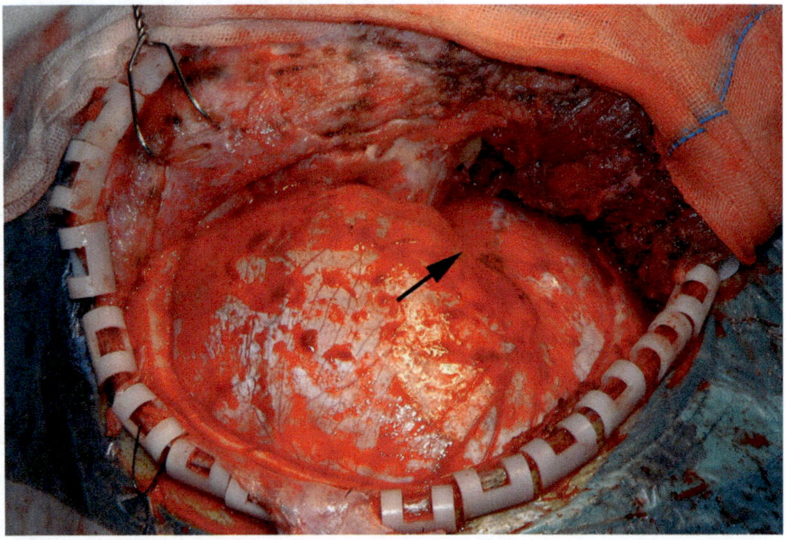

Fig. 29 After flattening of sphenoid wing, frontotemporal dura, and Sylvian fissure (arrow) were seen

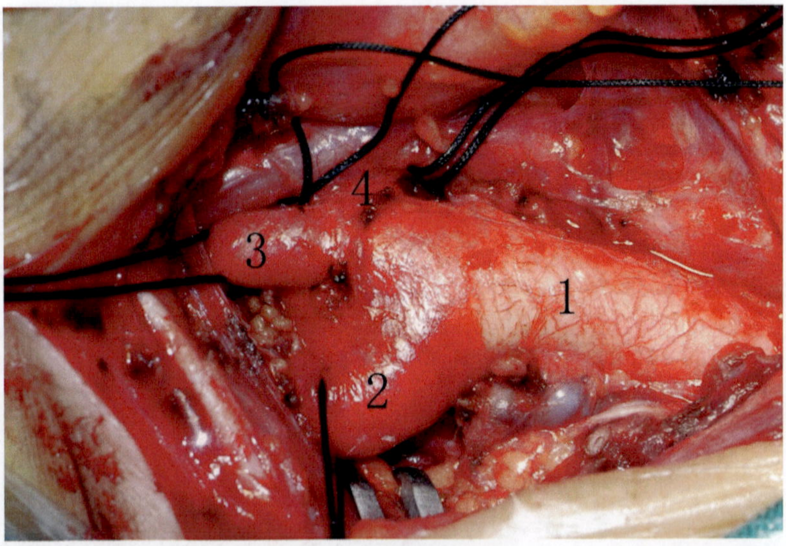

Fig. 30 At the same time, cervical carotid complex was exposed for proximal occlusion if an unexpected rupture was encountered. (*1*) Common carotid artery, (*2*) internal carotid artery, (*3*) external carotid artery, and (*4*) superior thyroid artery

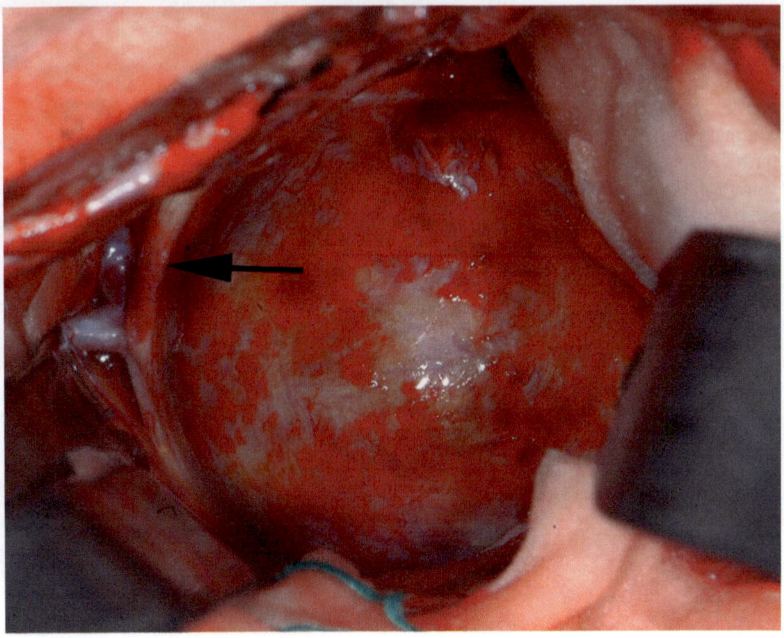

Fig. 31 The aneurysm was giant and thrombosed. The right optic nerve (arrow) was distorted by the lesion

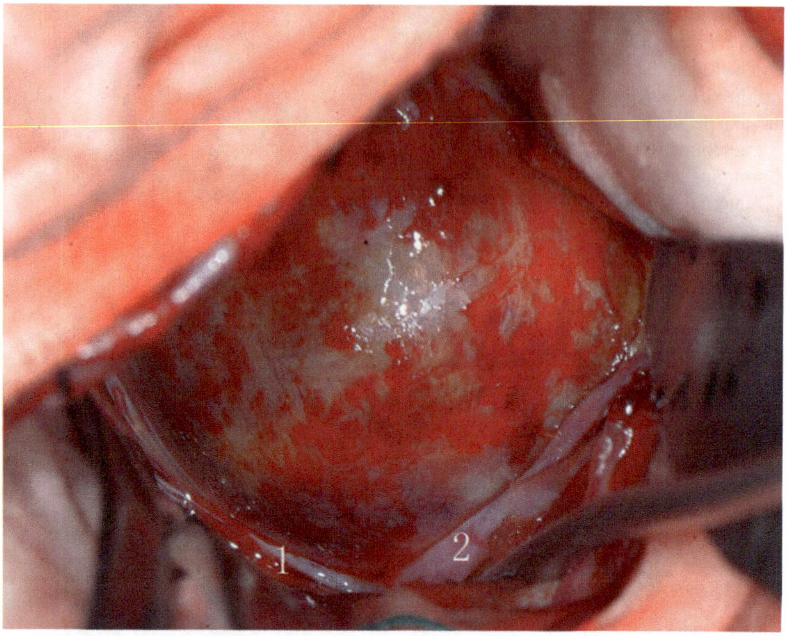

Fig. 32 The aneurysm was further dissected away from outflow arteries. (*1*) Precommunicating segment of anterior cerebral artery and (*2*) sphenoidal segment of middle cerebral artery

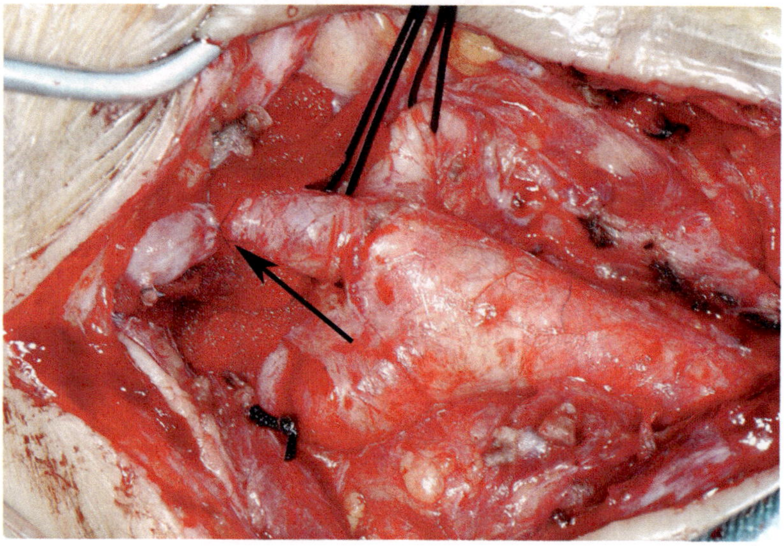

Fig. 33 Following external carotid artery (arrow)-to-saphenous vein-to-middle cerebral artery bypass, cervical internal carotid artery was ligated

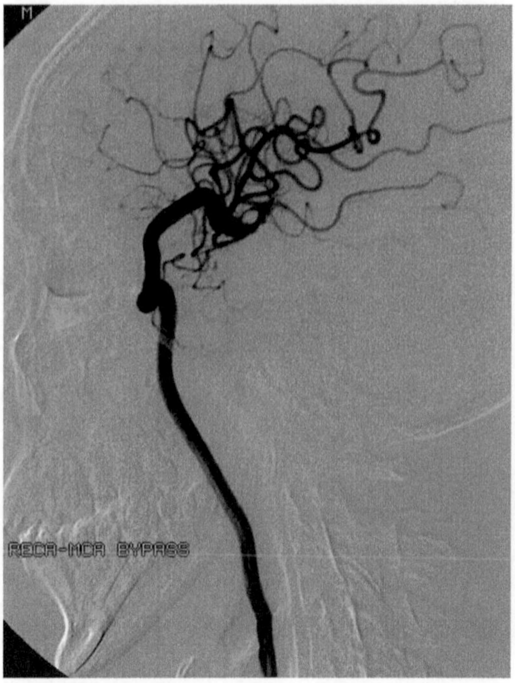

Fig. 34 Post-op lateral view of right external carotid artery angiography

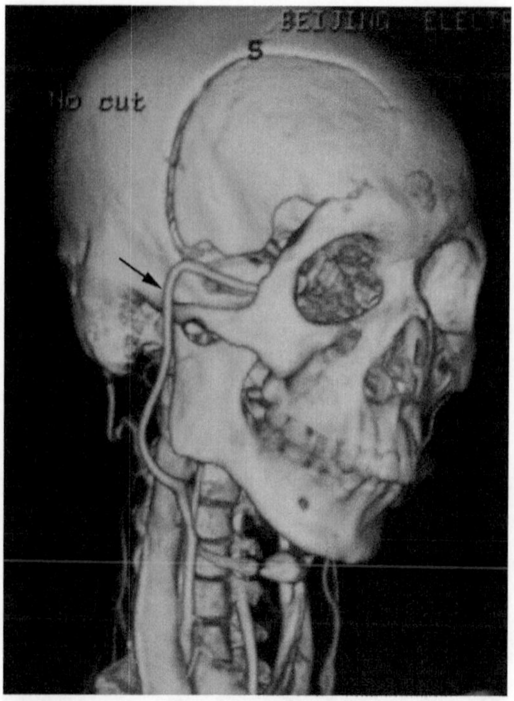

Fig. 36 3D CT angiography of external status at 1-year follow-up

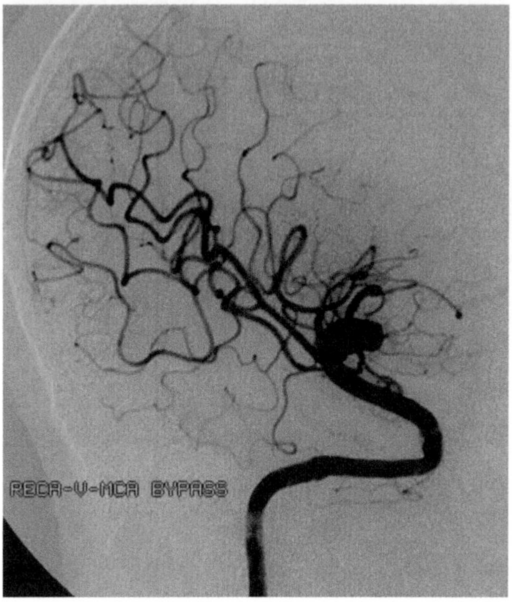

Fig. 35 Post-op lateral view of right external carotid artery angiography. Post-op right external carotid artery angiography confirmed patent status of graft with decreased volume of aneurysm

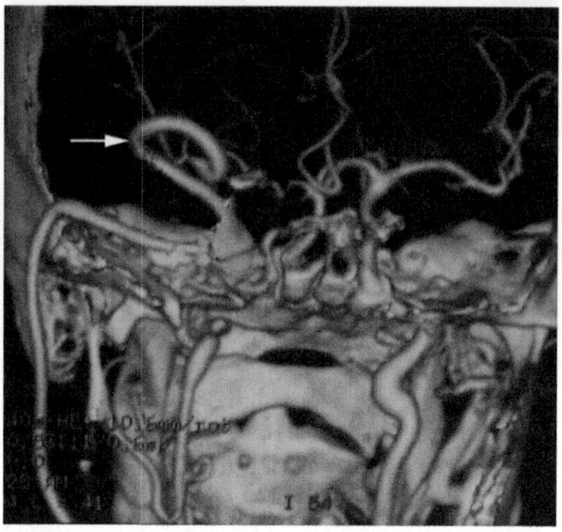

Fig. 37 3D CT angiography of internal status at 1-year follow-up. 3D CT angiography at 1-year follow-up showed disappearance of aneurysm with graft patency

Case 4.3 Basilar Trunk Aneurysm

Clinical presentation:

A 35-year-old female presented with a history of sudden onset of headache for 4 days.

Neurological exam:

Unremarkable

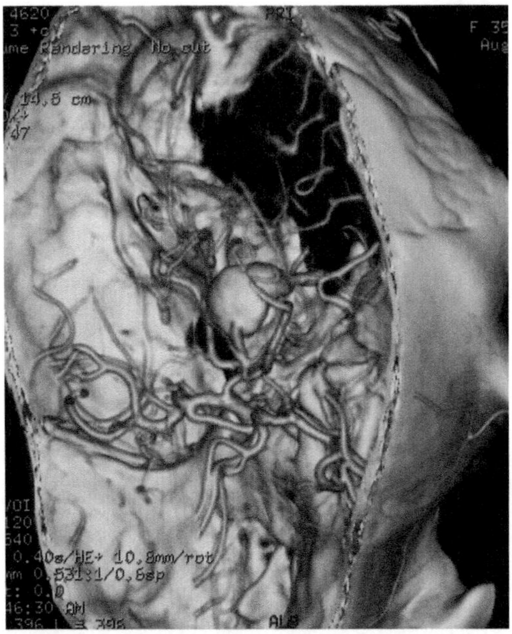

Fig. 38 Pre-op lateral view of 3D CT angiography

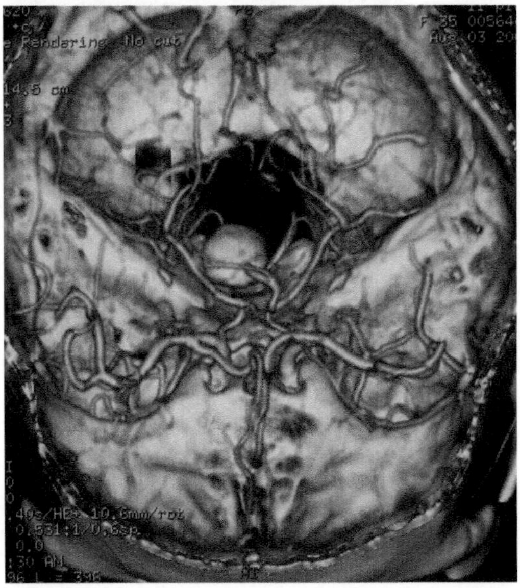

Fig. 39 Pre-op superior view of 3D CT angiography. Pre-op 3D CT angiography revealed a giant basilar trunk fusiform aneurysm measuring 3 × 3 × 3 cm

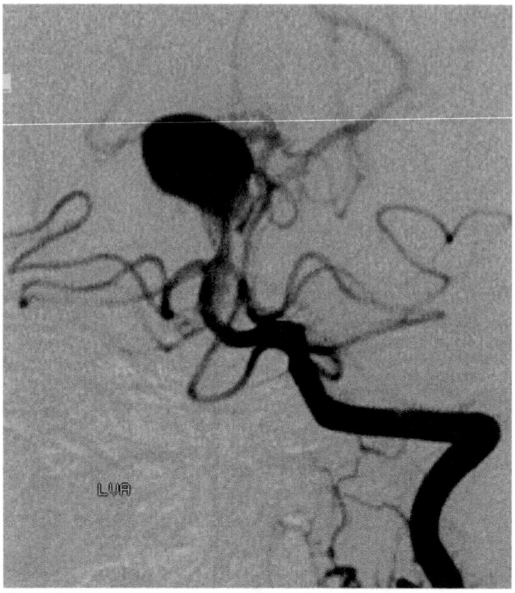

Fig. 40 Pre-op oblique view of left vertebrobasilar artery angiography

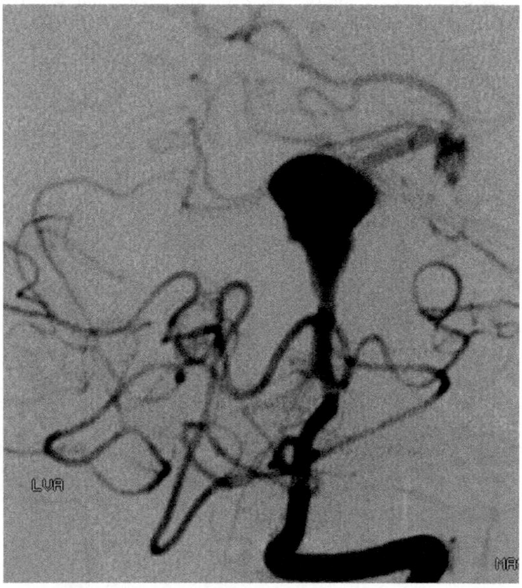

Fig. 41 Pre-op anteroposterior view of left vertebrobasilar artery angiography. Pre-op left vertebrobasilar artery angiography confirmed aneurysmal dilation of basilar trunk

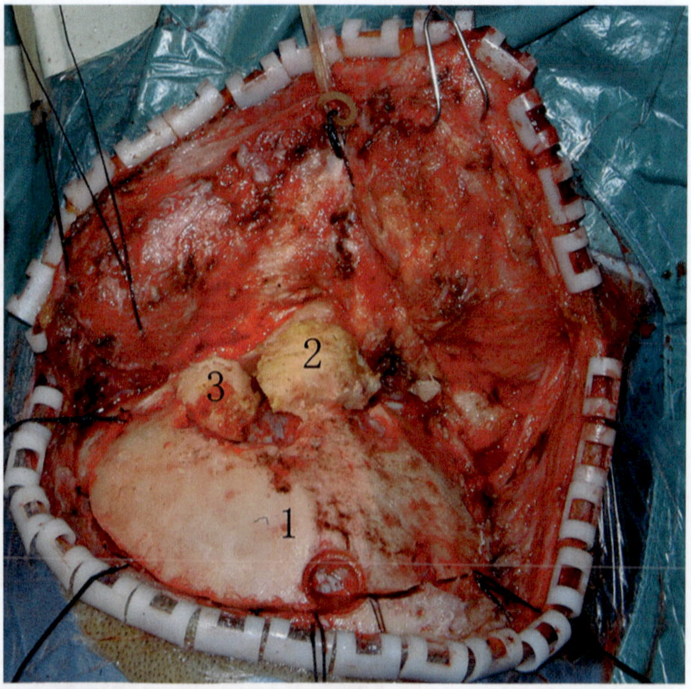

Fig. 42 The combined pre-sigmoid and retrosigmoid craniotomy was performed and bony flap was removed in 3 pieces. (*1*) main bone flap, (*2*) mastoid part of temporal bone, and (*3*) squamous part of temporal bone

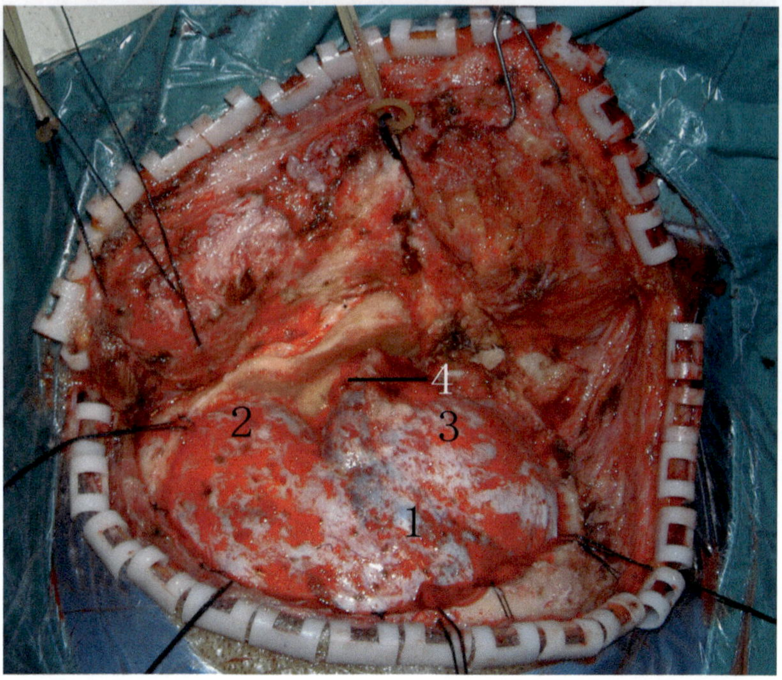

Fig. 43 (*1*) Transverse sinus; (*2*) middle fossa dura; (*3*) posterior fossa dura; and (*4*) sigmoid sinus

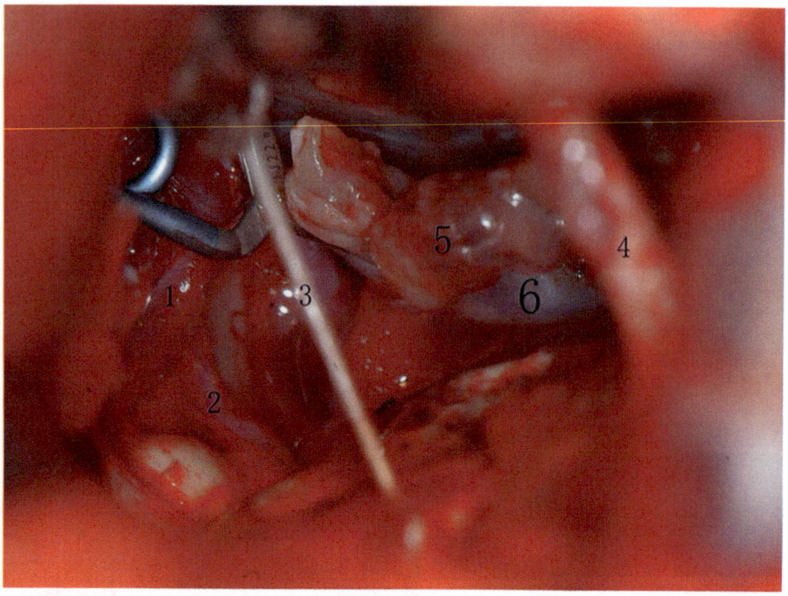

Fig. 44 A curved aneurysm clip (FT786T) was initially used to repair the distal portion of aneurysm neck by upside-down fashion to release to blades. (*1*) Ipsilateral posterior cerebral artery, (*2*) ipsilateral superior cerebellar artery, (*3*) abducens nerve, (*4*) trigeminal nerve, (*5*) aneurysm, and (*6*) basilar trunk

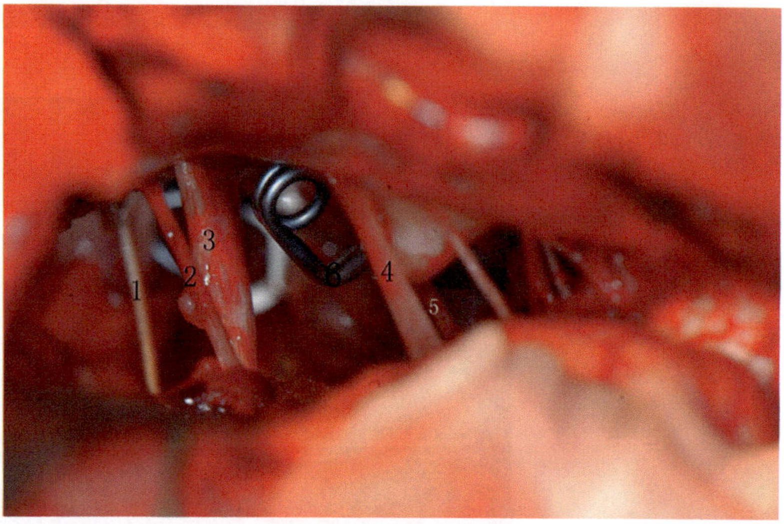

Fig. 45 A straight clip was added to close the residual portion of proximal neck in a face-to-face manner. (*1*) Trochlear nerve, (*2*) cochlear nerve, (*3*) facial nerve, (*4*) glossopharyngeal nerve, (*5*) hypoglossal nerve, and (*6*) clip

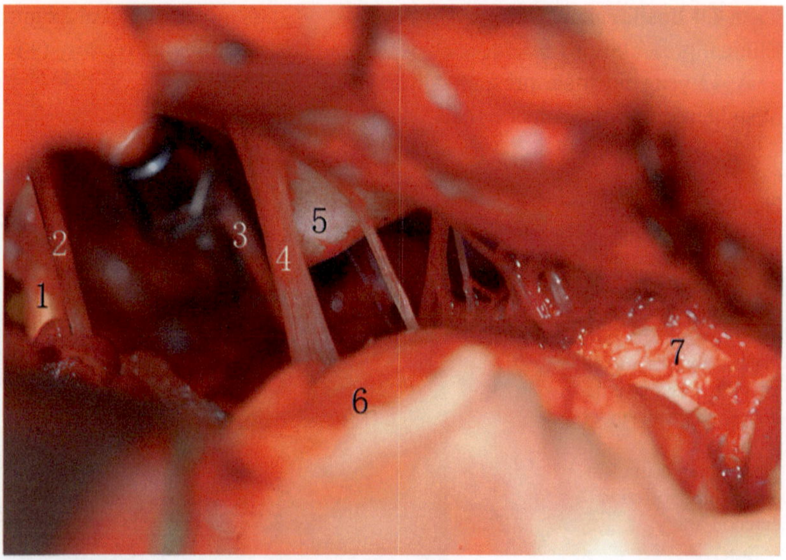

Fig. 46 The following structures were demonstrated in the surgical field following repair of aneurysm. (*1*) Facial nerve, (*2*) cochlear nerve, (*3*) abducens nerve, (*4*) glossopharyngeal nerve, (*5*) jugular tubercle, (*6*) inferior semilunar lobule, and (*7*) medulla oblongata

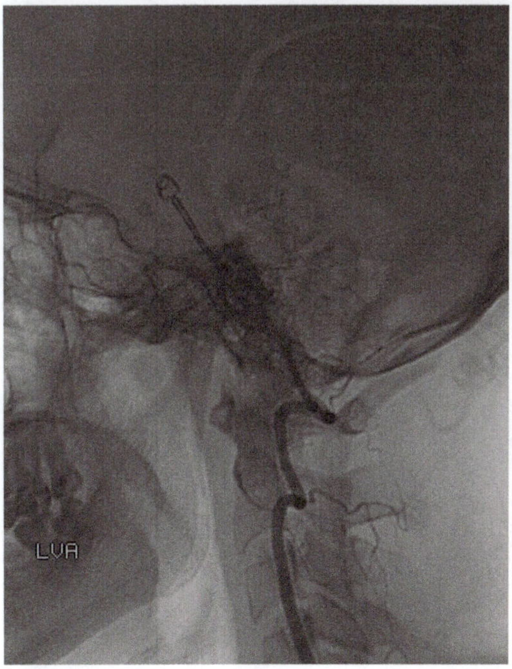

Fig. 47 Post-op lateral view of left vertebrobasilar artery angiography

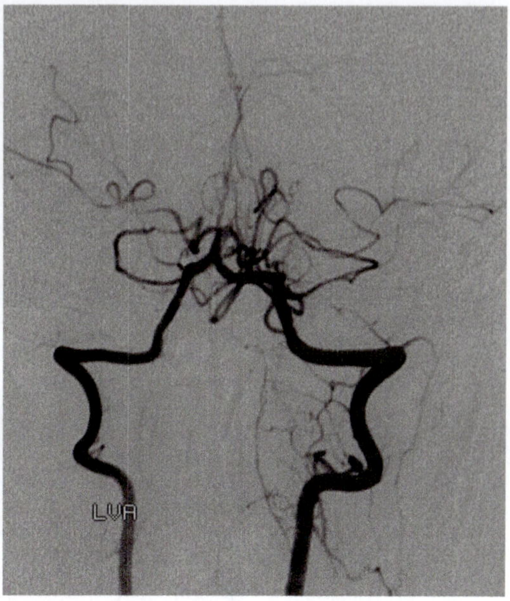

Fig. 48 Post-op anteroposterior view of left vertebrobasilar artery angiography. Post-op left vertebrobasilar artery angiography showed obliteration of aneurysm

Case 4.4 Basilar Apex Aneurysm

Clinical presentation:

A 42-year-old male presented with a history of sudden onset of headache for 18 days.

Neurological exam:

Nuchal rigidity (+)

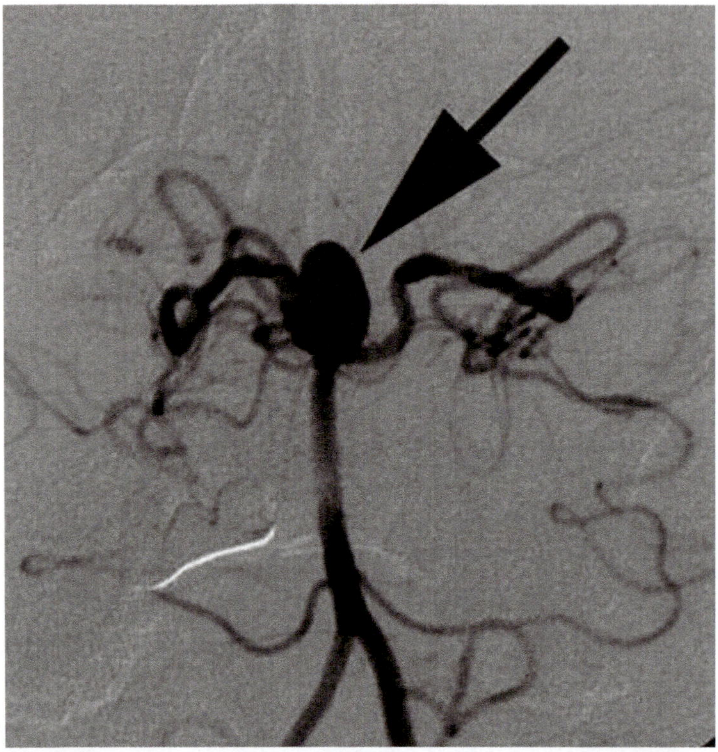

Fig. 49 Pre-op anteroposterior view of vertebrobasilar artery angiography revealed a saccular basilar tip aneurysm with posterior thalamoperforating artery arising from the dome and involvement of bilateral precommunicating segment of posterior cerebral artery

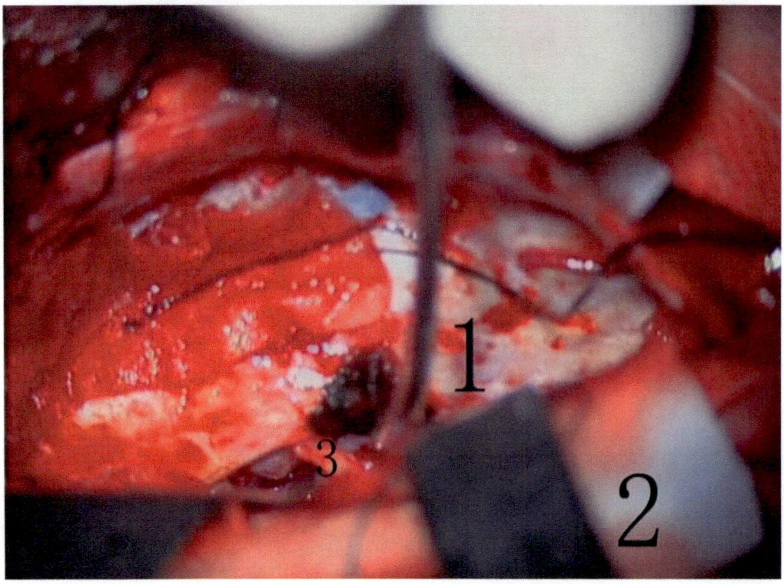

Fig. 50 The right intradural subtemporal approach was chosen to reach the nidus. Tentorium was incised at a site where 0.5 cm posterior to the 4th nerve penetration into the notch. (*1*) Tentorium, (*2*) temporal lobe, and (*3*) oculomotor nerve

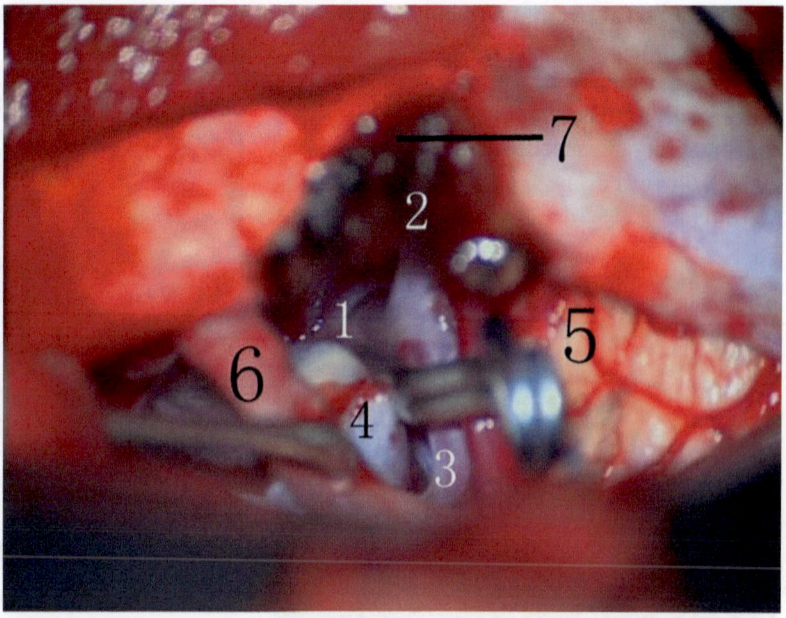

Fig. 51 The aneurysm was repaired directly using a straight clip. (*1*) Basilar trunk, (*2*) right posterior cerebral artery, (*3*) left posterior cerebral artery, (*4*) aneurysm, (*5*) pons, (*6*) oculomotor nerve, and (*7*) tentorial notch

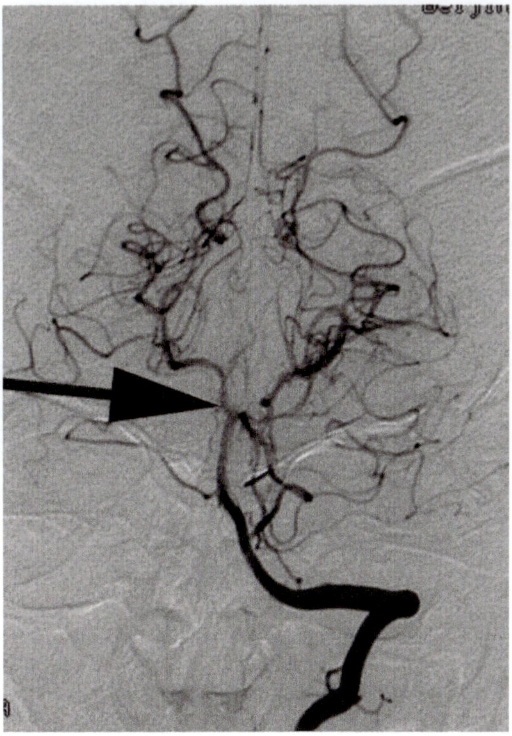

Fig. 52 Post-op anteroposterior view of left vertebrobasilar artery angiography

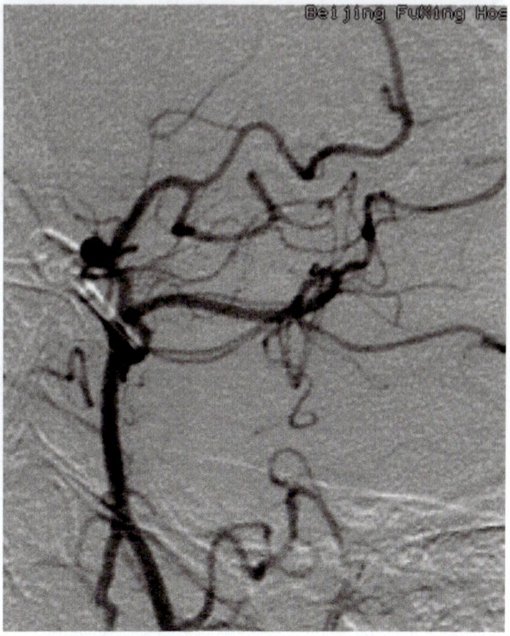

Fig. 53 Post-op lateral view of left vertebrobasilar artery angiography. Post-op left vertebrobasilar artery angiography demonstrated no residual aneurysm with preservation of perforating arteries

Case 4.5 Lateral Frontal Arteriovenous Malformation

Clinical presentation:

30-year-old male presented with a history of sudden onset of headache for one week.

Neurological exam:

Unremarkable

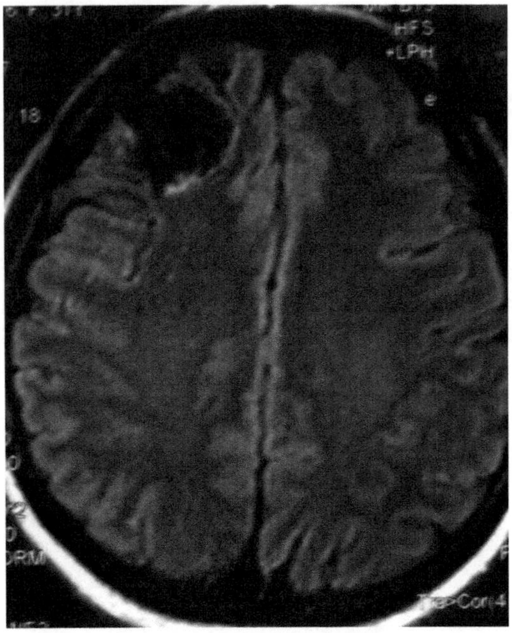

Fig. 54 Pre-op axial T1-weighted MRI

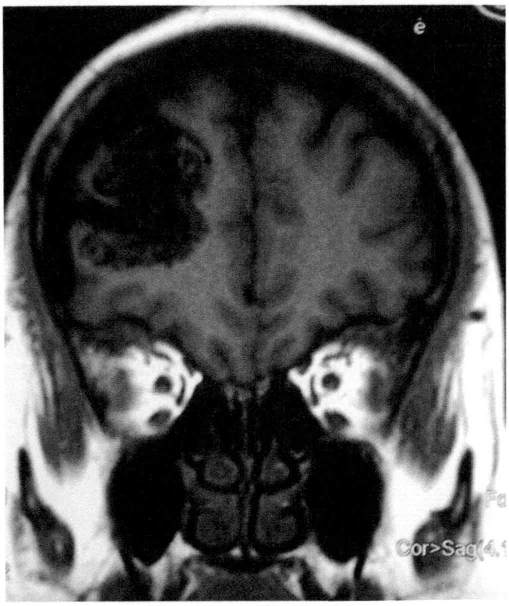

Fig. 55 Pre-op coronal T1-weighted MRI

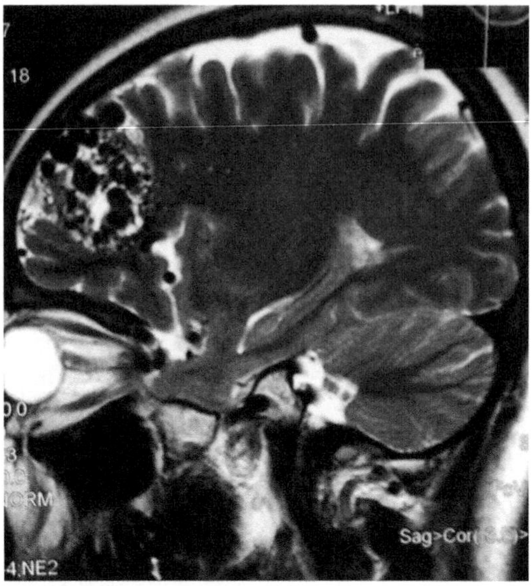

Fig. 56 Pre-op sagittal T2-weighted MRI. Pre-op brain MRI demonstrated large dilated serpiginous and clustered nidiform flow voids centered on the right frontal lobe

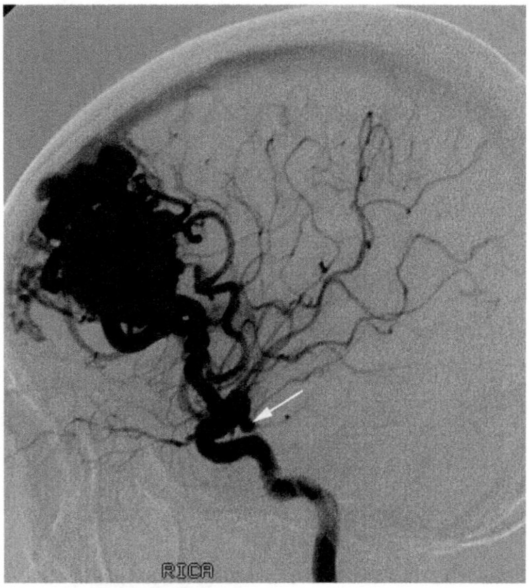

Fig. 57 Pre-op lateral view of right internal carotid artery angiography showed a lateral frontal arteriovenous malformation fed by frontopolar artery (branch of anterior cerebral artery) and prefrontal artery (branch of middle cerebral artery). A hemodynamic-related aneurysm (arrow) of posterior communicating segment of internal carotid artery was seen

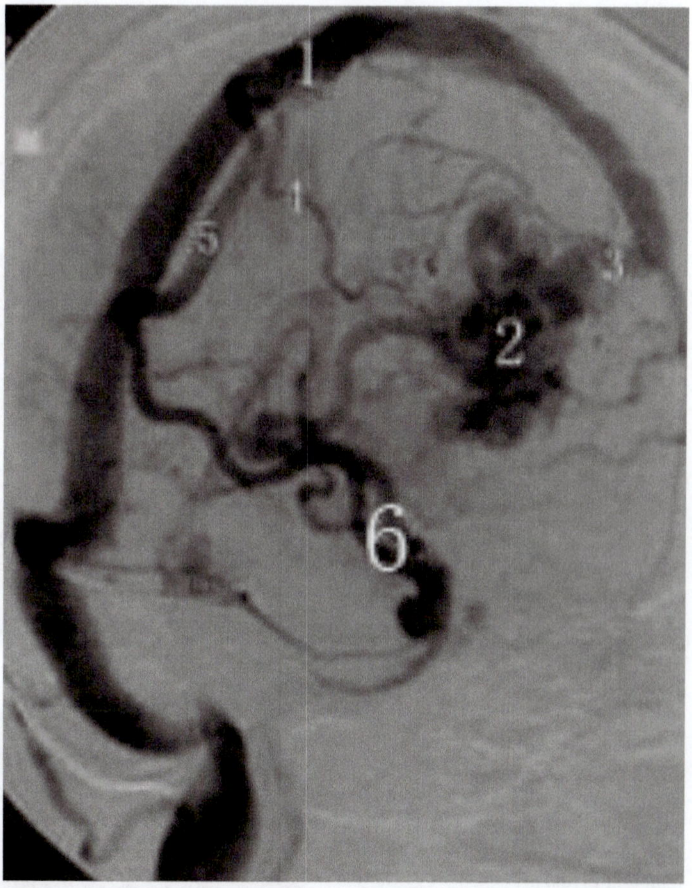

Fig. 58 Pre-op oblique view of venous phase of right internal carotid artery angiography showed a lateral frontal arteriovenous malformation was drained anteriorly by frontopolar vein to superior sagittal sinus and posteriorly by superficial Sylvian vein to the vein of Labbe. (*1*) Superior sagittal sinus, (*2*) AVM nidus, (*3*) frontopolar vein, (*4*) postfrontal vein, (*5*) vein of Trolend, and (*6*) vein of Labbe

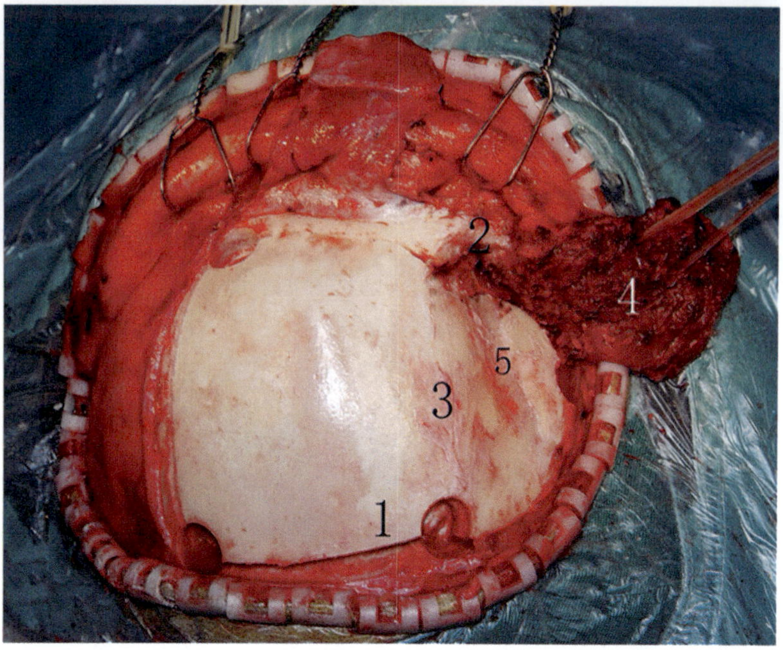

Fig. 59 The right frontotemporal craniotomy was carried out. (*1*) Coronal suture, (*2*) zygomatic process of frontal bone, (*3*) superior temporal line, (*4*) temporalis, and (*5*) pterion

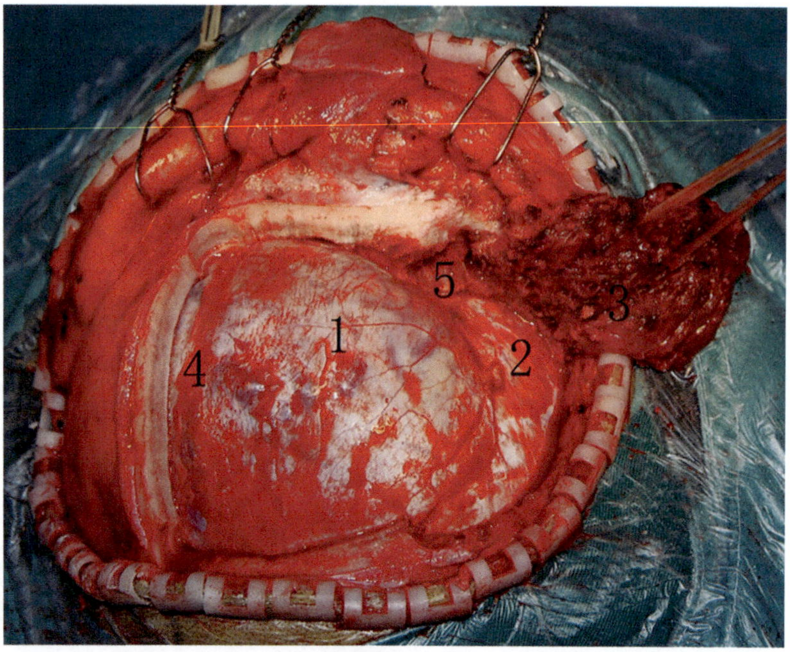

Fig. 60 Frontotemporal dura was unveiled after bony removal. (*1*) Frontal dura, (*2*) temporal dura, (*3*) temporalis, (*4*) superior sagittal sinus, and (*5*) sphenoid ridge

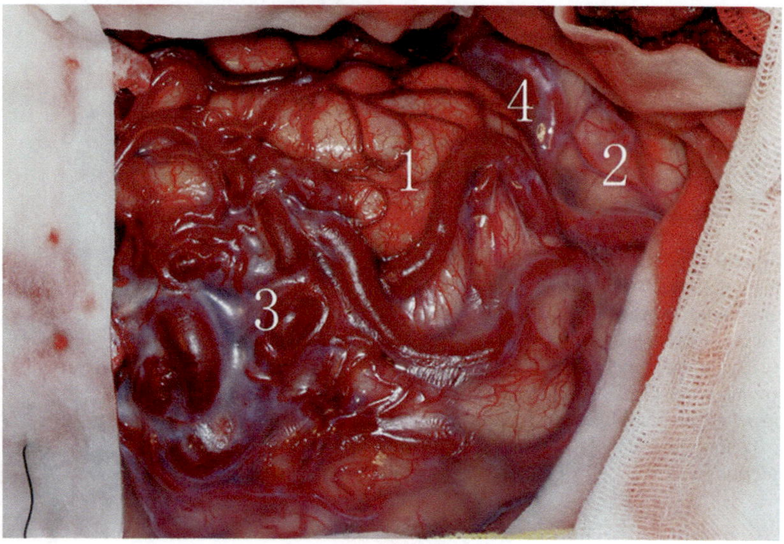

Fig. 61 The serpiginous lesion was exposed at the lateral surface of frontal lobe. (*1*) Pars triangularis; (*2*) anterior portion of superior temporal gyrus; (*3*) arteriovenous malformation; and (*4*) Sylvian fissure

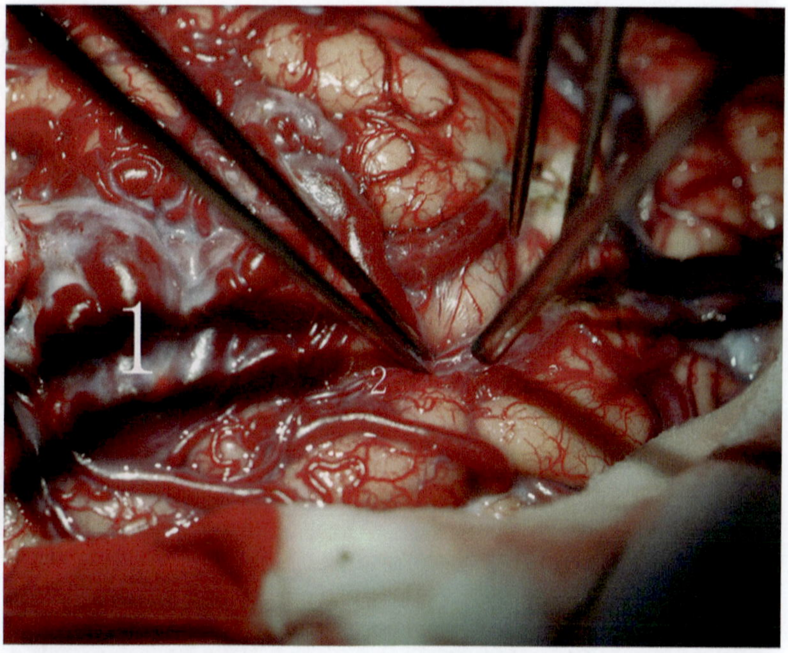

Fig. 62 The feeders were dissected, coagulated, and cut along the boundary of nidus. (*1*) Arteriovenous malformation and (*2*) prefrontal artery

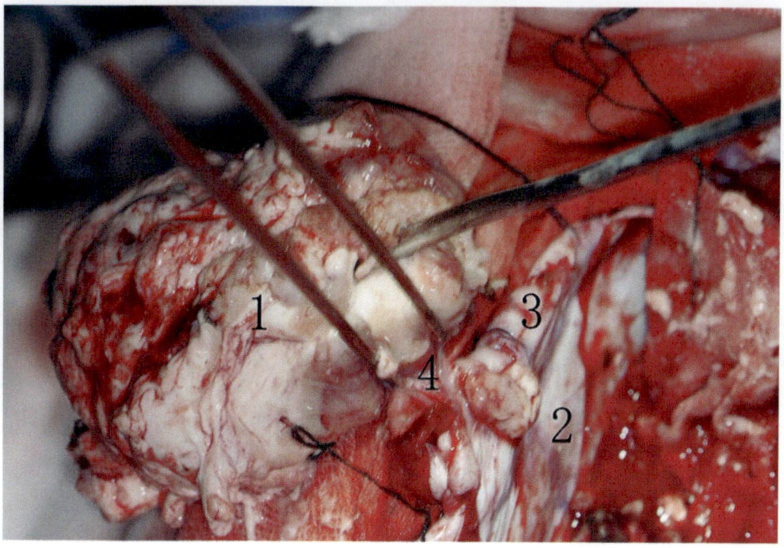

Fig. 63 After circumferential dissection of lesion, the main draining vein was cut. (*1*) Nidus, (*2*) falx, (*3*) superior sagittal sinus, and (*4*) main draining vein

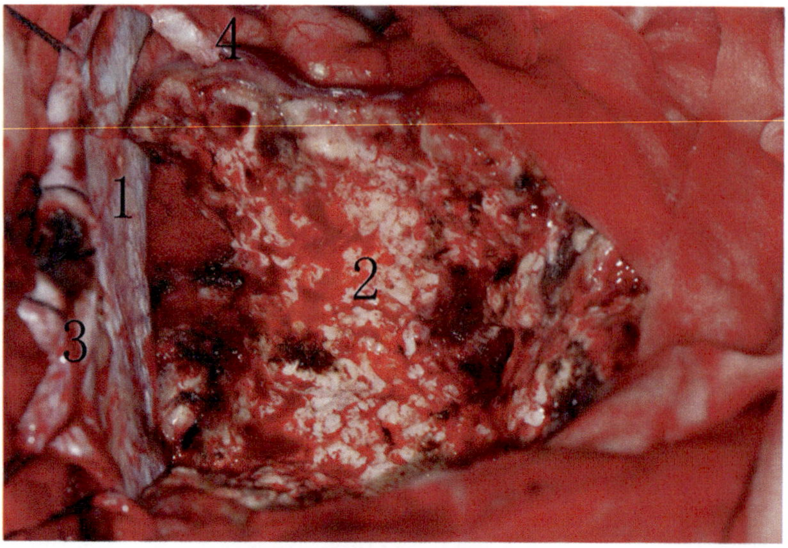

Fig. 64 Surgical bed was meticulously hemostased. (*1*) Falx, (*2*) frontal lobe, (*3*) superior sagittal sinus, and (*4*) frontal pole

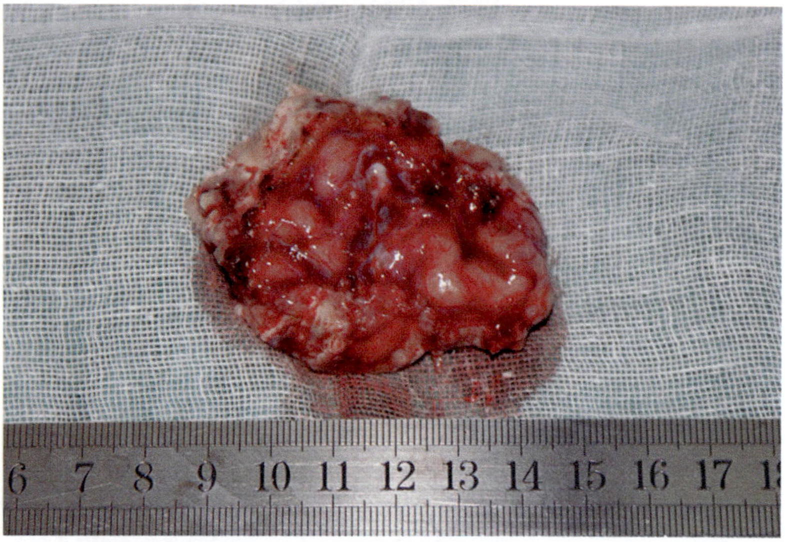

Fig. 65 The AVM specimen was measured at 5 cm

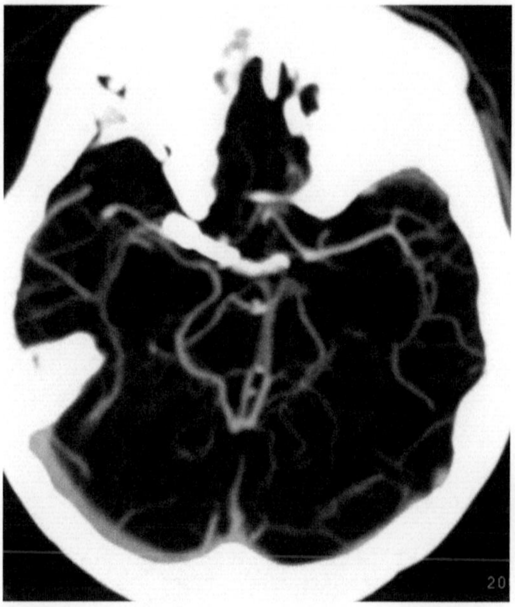

Fig. 66 Post-op axial view of 3D CT angiography

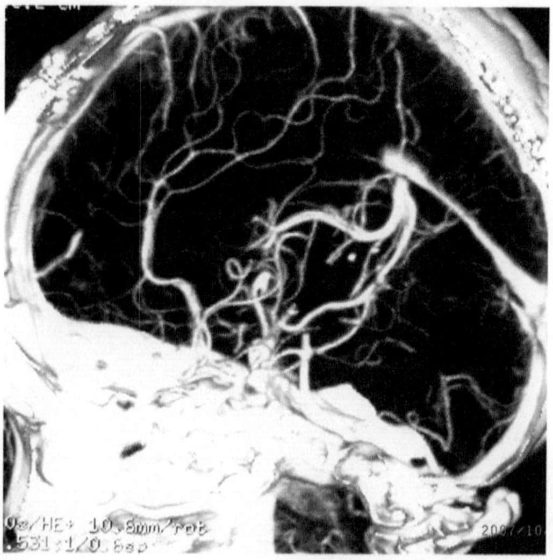

Fig. 67 Post-op sagittal view of 3D CT angiography. Post-op 3D CT angiography demonstrated total resection of AVM

Anatomy of Tentorial Incisure, Ambient Cistern and Surgical Techniques

Xiang'En Shi, Long Wang, and Hai Qian

X. Shi (✉) · L. Wang · H. Qian
Neurological Surgery, SanBo Brain Hospital, Capital Medical University, Beijing, China
e-mail: shixen@ccmu.edu.cn;
wanglong@ccmu.edu.cn; qianh@ccmu.edu.cn

X. Shi et al. (eds.), *Microneurosurgical Anatomy and Surgical Technique*,
https://doi.org/10.1007/978-981-19-8273-6_5

1 Anatomy of Tentorial Incisure

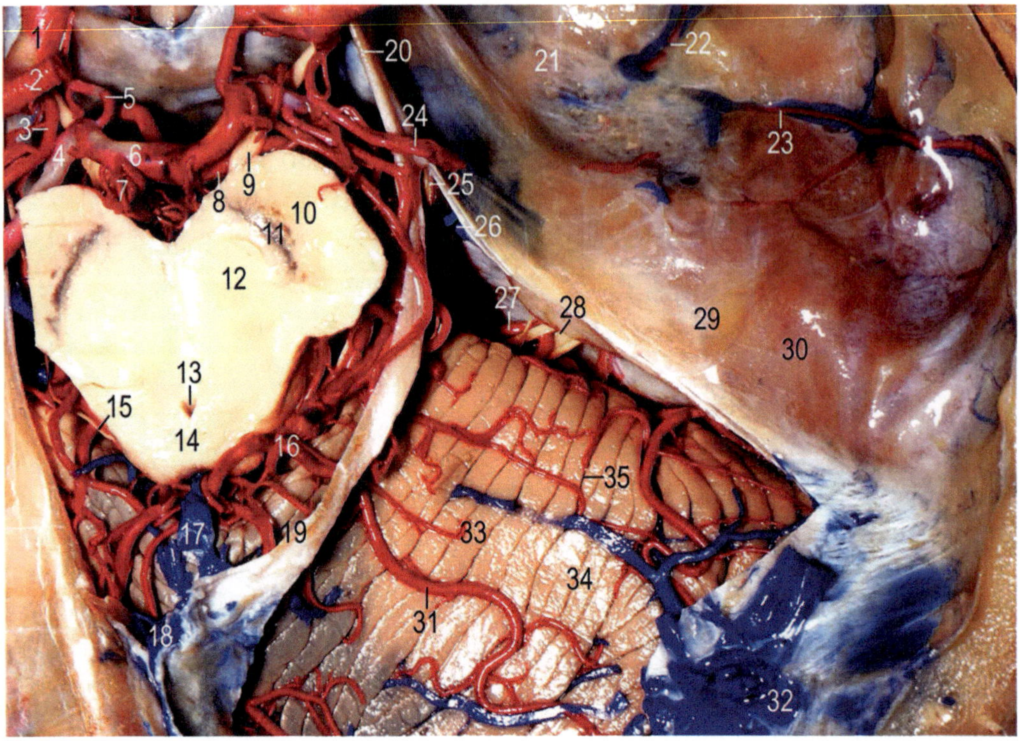

Fig. 1 Superior view of right tentorial incisura, midbrain, and tentorial surface of cerebellum. (*1*) Internal carotid artery, (*2*) Anterior cerebral artery, (*3*) Anterior choroidal artery, (*4*) Postcommunicating P2 segment of posterior cerebral artery, (*5*) Posterior communicating artery, (*6*) Precommunicating P1 segment of posterior cerebral artery, (*7*) Posterior thalamoperforating arteries perforating artery, (*8*) Superior cerebellar artery, (*9*) Oculomotor nerve, (*10*) Cerebral peduncle, (*11*) Substantia nigra of midbrain, (*12*) Red nucleus, (*13*) Cerebral aqueduct of midbrain, (*14*) Tectum of midbrain, (*15*) Trochlear nerve, (*16*) Proximal hemispheric branches of superior cerebellar artery, (*17*) Vein of Galen, (*18*) Straight sinus rectus, (*19*) Tentorial Tentorium edge, (*20*) Anterior petroclinoid fold, (*21*) Trigeminal ganglion, (*22*) Frontal branch of middle meningeal artery, (*23*) Petrosqusamosal branch of middle meningeal artery, (*24*) Anterior choroid artery, (*25*) Trigeminal nerve, (*26*) Superior petrosal vein, (*27*) Anterior inferior cerebellar artery, (*28*) Facial-acoustic nerve complex, (*29*) Arcuate eminence, (*30*) Tegmen tympani, (*31*) Distal hemispheric branches of superior cerebellar artery, (*32*) Lateral tentorial tentorium sinus, (*33*) Quadrangular lobule, (*34*) Simple lobule, (*35*) Primary fissure

Fig. 2 Enlarged superior view of anterior and middle portions of right tentorial incisura. (*1*) Pituitary gland, (*2*) Pituitary stalk, (*3*) Dorsum sellae, (*4*) Optic nerve, (*5*) Internal carotid artery, (*6*) Anterior cerebral artery, (*7*) Middle cerebral artery, (*8*) Posterior communicating artery, (*9*) Precommunicating segment of posterior cerebral artery, (*10*) Postcommunicating segment of posterior cerebral artery, (*11*) Anterior clinoid process, (*12*) Superior wall of cavernous sinus, (*13*) Anterior petroclinoid fold, (*14*) Lateral wall of cavernous sinus, (*15*) Oculomotor nerve, (*16*) Clivus, (*17*) Trochlear nerve, (*18*) Abducens nerve, (*19*) Cerebral peduncle, (*20*) Substantia nigra, (*21*) Optic strut, (*22*) Posterior petroclinoid fold

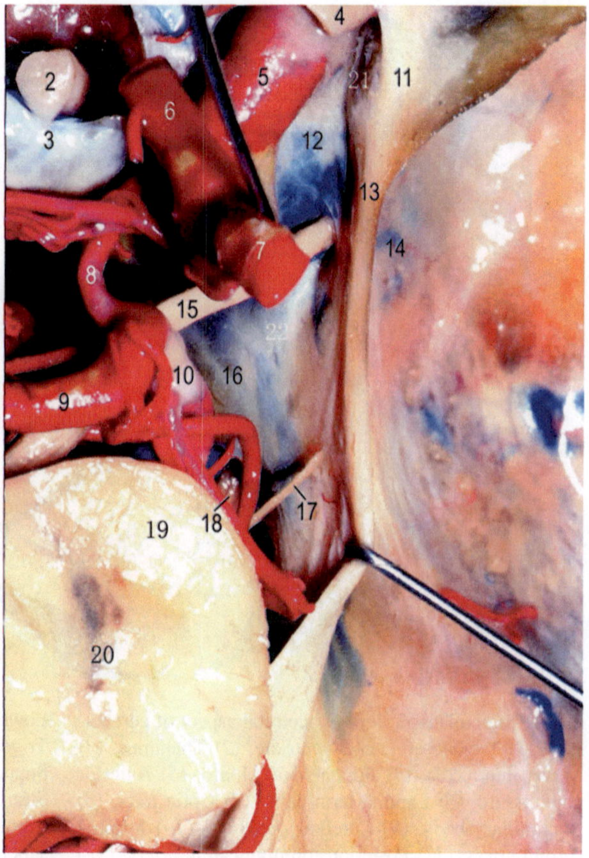

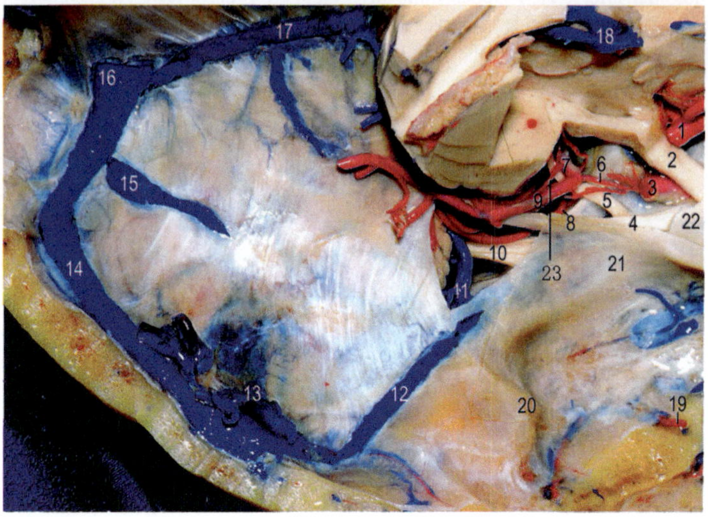

Fig. 3 Superior view of right tentorium. (*1*) Anterior cerebral artery, (*2*) Optic nerve, (*3*) Internal carotid artery, (*4*) Anterior petroclinoid fold, (*5*) Oculomotor nerve, (*6*) Posterior communicating artery, (*7*) Precommunicating segment of posterior cerebral artery, (*8*) Superior cerebellar artery, (*9*) Postcommunicating segment of posterior cerebral artery, (*10*) Trigeminal nerve, (*11*) Superior petrosal vein, (*12*) Superior petrosal sinus, (*13*) Lateral tentorial sinus, (*14*) Transverse sinus, (*15*) Medial tentorial sinus, (*16*) Confluence of sinus, (*17*) Straight sinus, (*18*) Internal cerebral vein, (*19*) Middle meningeal artery, (*20*) Arcuate eminence, (*21*) Lateral wall of cavernous sinus, (*22*) Anterior clinoid process, (*23*) Peduncular perforating arteries

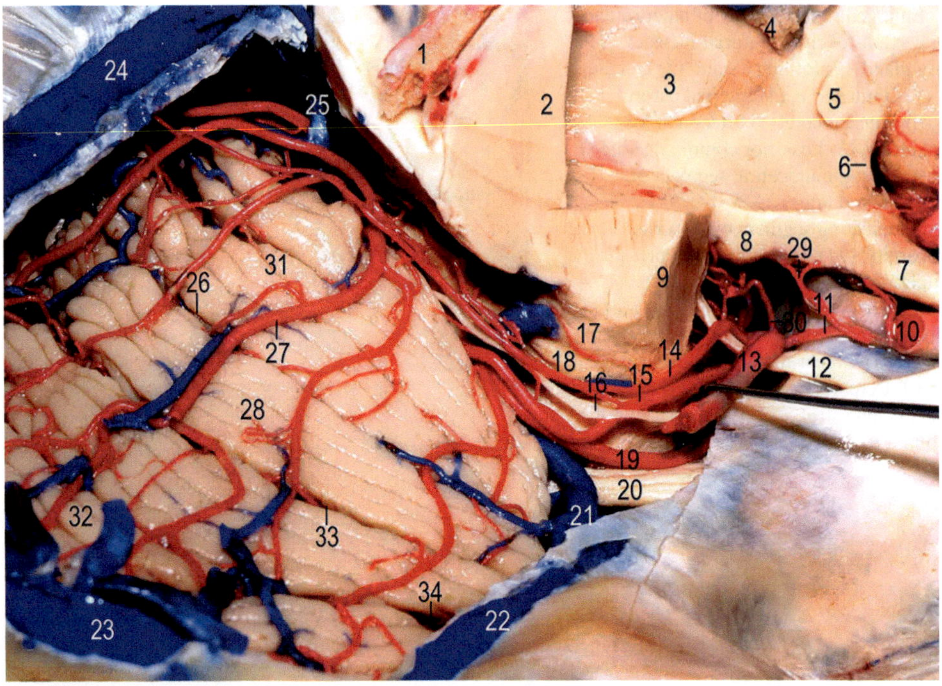

Fig. 4 Superolateral view of occipital aspect of cerebellum. (*1*) Trigone of lateral ventricle, (*2*) Thalamus, (*3*) Massa intermedia, (*4*) Foramen of Monro, (*5*) Anterior commissure, (*6*) Lamina terminalis, (*7*) Optic tract, (*8*) Mammillary body, (*9*) Cerebral peduncle, (*10*) Internal carotid artery, (*11*) Posterior communicating artery, (*12*) Oculomotor nerve, (*13*) Posterior cerebral artery, (*14*) Medial posterior choroidal artery, (*15*) Upper trunk of superior cerebellar artery, (*16*) Trochlear nerve, (*17*) Lateral geniculate body, (*18*) Medial geniculate body, (*19*) Lower Inferior trunk of superior cerebellar artery, (*20*) Trigeminal nerve, (*21*) Superior petrosal vein, (*22*) Superior petrosal sinus, (*23*) Transverse sinus, (*24*) Straight sinus rectus, (*25*) Superior cerebellar vein, (*26*) Primary (Tentorial) fissure, (*27*) Hemispheric branch of superior cerebellar artery, (*28*) Posterior part of quadrangular lobule, (*29*) Premamillary artery, (*30*) Posterior thalamoperforating arteries, (*31*) Anterior part of quadrangular lobule, (*32*) Superior semilunar lobule, (*33*) Simple lobule, (*34*) Postclival fissure

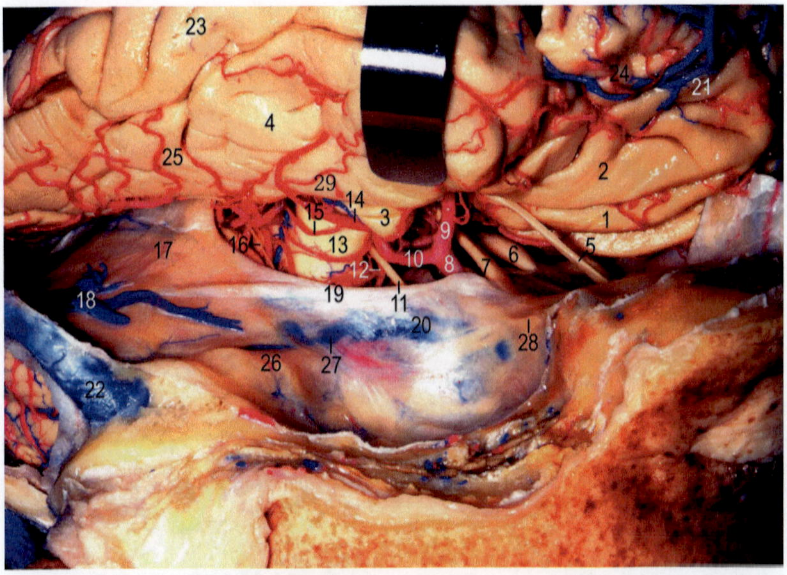

Fig. 5 Surgical overview of subtemporal approach. (*1*) Gyrus rectus, (*2*) Medial orbital gyrus, (*3*) Uncus, (*4*) Lateral Occipitotemporal gyrus, (*5*) Olfactory nerve, (*6*) Left optic nerve, (*7*) Right optic nerve, (*8*) Ophthalmic segment of internal carotid artery, (*9*) Postcommunicating segment of internal carotid artery, (*10*) Posterior communicating artery, (*11*) Oculomotor nerve, (*12*) Superior cerebellar artery, (*13*) Mesencephalon, (*14*) Posterior cerebral artery, (*15*) Medial posterior choroidal artery, (*16*) Trochlear nerve, (*17*) Tentorium of cerebellum, (*18*) Lateral tentorial sinus, (*19*) Tentorial incisure, (*20*) Lateral wall of cavernous sinus, (*21*) Anterior orbital gyrus, (*22*) Sigmoid sinus, (*23*) Inferior temporal gyrus, (*24*) Lateral orbital gyrus, (*25*) Fusiform gyrus, (*26*) Arcuate eminence, (*27*) Superior petrosal sinus, (*28*) Sphenoid ridge, (*29*) Parahippocampal gyrus

2 Anatomy of Ambient Cistern

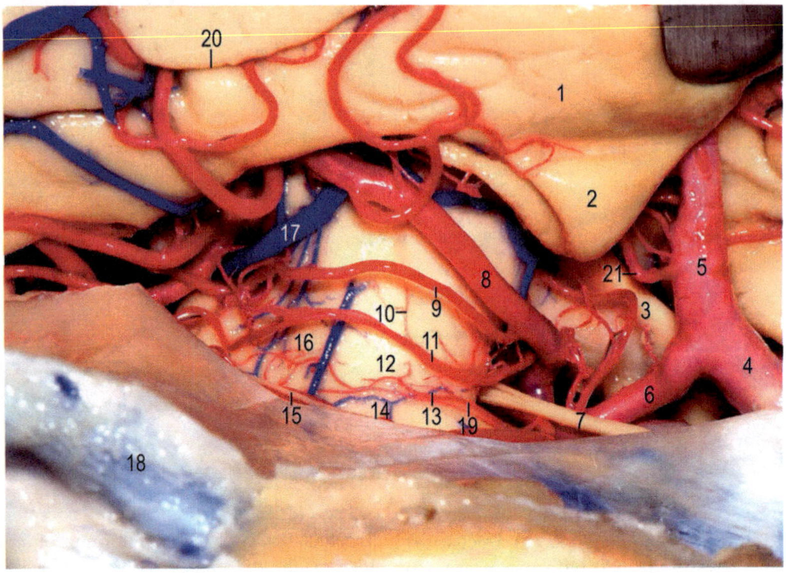

Fig. 6 Subtemporal approach to ambient cistern and lateral pontine region. (*1*) Parahippocampal gyrus, (*2*) Uncus, (*3*) Optic tract, (*4*) Ophthalmic segment of internal carotid artery, (*5*) Postcommunicating segment of internal carotid artery, (*6*) Posterior communicating artery, (*7*) Oculomotor nerve, (*8*) Posterior cerebral artery, (*9*) Lateral posterior choroidal artery, (*10*) Short circumflex arteries, (*11*) Medial posterior choroidal artery, (*12*) Midbrain, (*13*) Pontomesencephalic sulcus, (*14*) Pons, (*15*) Trochlear nerve, (*16*) Tegmental surface, (*17*) Basal vein of Rosenthal, (*18*) Transverse sinus, (*19*) Long circumflex arteries, (*20*) Collateral sulcus, (*21*) Anterior choroidal artery

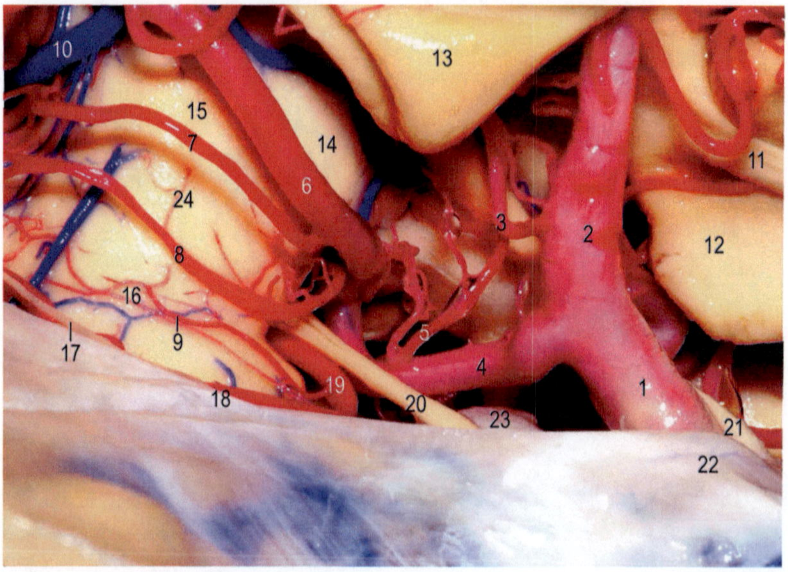

Fig. 7 Posterolateral view of ambient cistern. (*1*) Ophthalmic segment of internal carotid artery, (*2*) Postcommunicating segment of internal carotid artery, (*3*) Anterior choroidal artery, (*4*) Posterior communicating artery, (*5*) Premamillary artery, (*6*) Posterior cerebral artery, (*7*) Lateral posterior choroidal artery, (*8*) Medial posterior choroidal artery, (*9*) Long circumflex arteries, (*10*) Basal vein of Rosenthal, (*11*) Olfactory nerve, (*12*) Gyrus rectus, (*13*) Uncus, (*14*) Cerebral peduncle, (*15*) Midbrain, (*16*) Pontomesencephalic sulcus, (*17*) Trochlear nerve, (*18*) Tentorial edge, (*19*) Superior cerebellar artery, (*20*) Oculomotor nerve, (*21*) Optic nerve, (*22*) Anterior clinoid process, (*23*) Dorsum sellae, (*24*) Short circumflex arteries

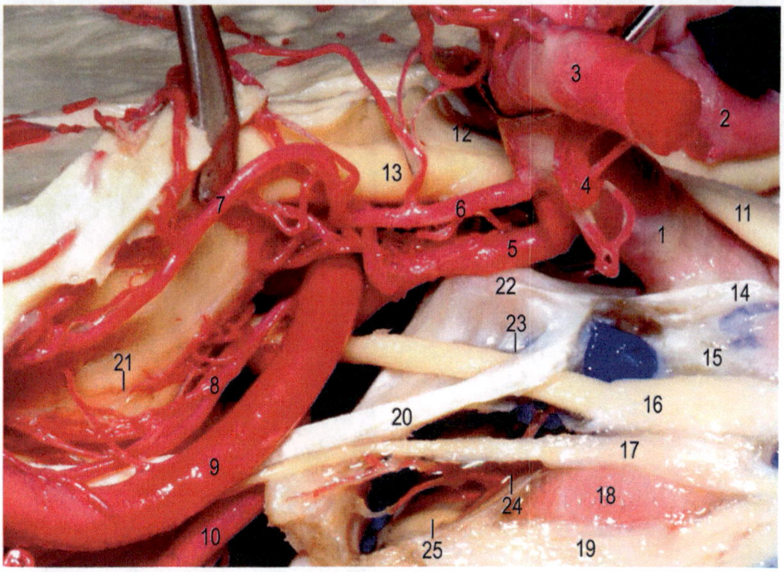

Fig. 8 Lateral view of anterior ambient cistern (*1*) Internal carotid artery, (*2*) Anterior cerebral artery, (*3*) Middle cerebral artery, (*4*) Lateral lenticulostriate artery, (*5*) Posterior communicating cerebral artery, (*6*) Cisternal segment of anterior choroidal artery, (*7*) Intraventricular segment of anterior choroidal artery, (*8*) Medial posterior choroidal artery, (*9*) Posterior cerebral artery, (*10*) Superior cerebellar artery, (*11*) Optic nerve, (*12*) Lamina terminalis, (*13*) Optic tract, (*14*) Upper (Distal) dural ring of ICA, (*15*) Lower (Proximal) dural ring Inferior ring of ICA, (*16*) Oculomotor nerve, (*17*) Trochlear nerve, (*18*) Cavernous segment of internal carotid artery ICA, (*19*) Ophthalmic nerve, (*20*) Anterior petroclinoid fold, (*21*) Inferior choroidal point, (*22*) Posterior clinoid process, (*23*) Oculomotor foramen, (*24*) Meningohypophyseal trunk, (*25*) Abducens nerve

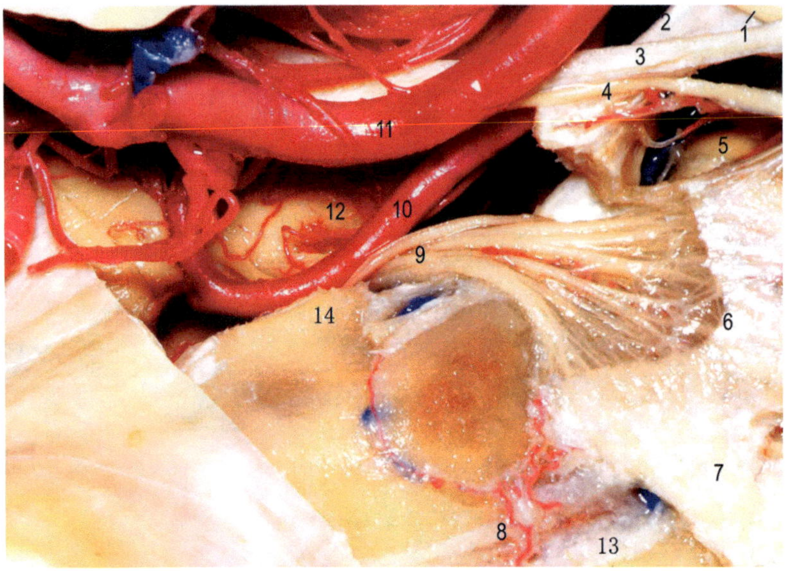

Fig 9 Lateral view of anterior ambient cistern with exposure of anterior petrous bone and Meckel's cave. (*1*) Oculomotor nerve, (*2*) Posterior petroclinoid fold, (*3*) Anterior petroclinoid fold, (*4*) Trochlear nerve, (*5*) Abducens nerve, (*6*) Trigeminal ganglion, (*7*) Mandibular nerve, (*8*) Greater superficial petrosal nerve, (*9*) Trigeminal root, (*10*) Superior cerebellar artery, (*11*) Posterior cerebral artery, (*12*) Cerebral peduncle, (*13*) Lesser superficial petrosal nerve, (*14*) Petrous apex

Case 5.1 Petroclival Meningioma

Clinical presentation:

A 43-year-old female with a 3-month history of dizziness and gait imbalance.

Neurological exam:

Finger-to-nose test (+), Romberg's sign (+)

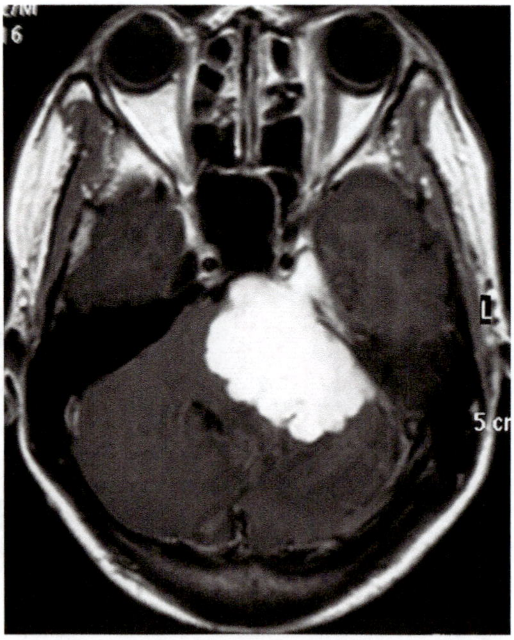

Fig. 10 Pre-op axial T1 Gd-enhanced MRI

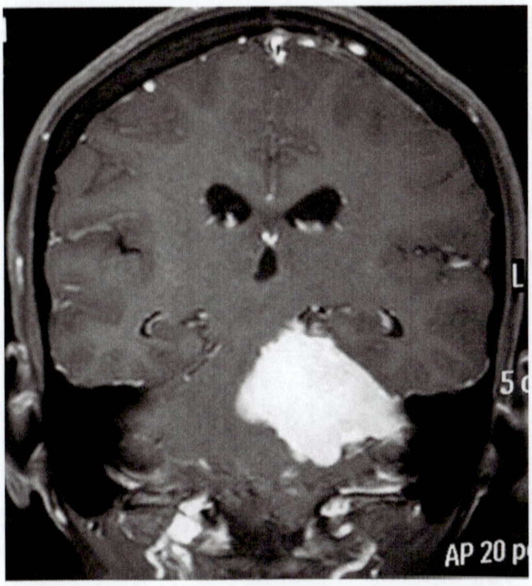

Fig. 11 Pre-op coronal T1 Gd-enhanced MRI

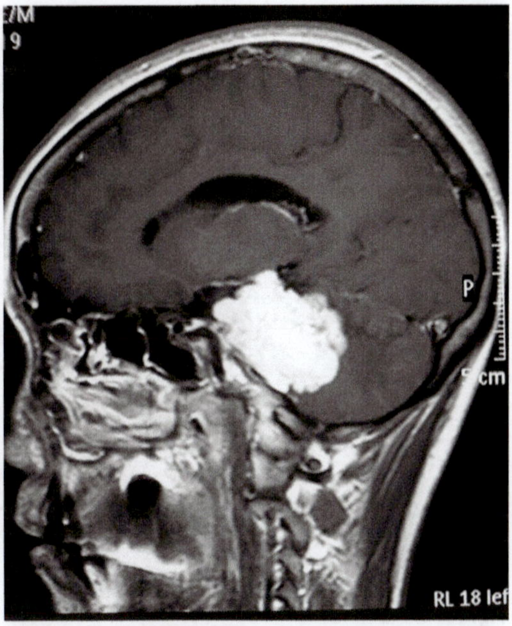

Fig. 12 Pre-op sagittal T1 Gd-enhanced MRI. Pre-op T1 Gd-enhanced MRI demonstrated a well-marginated petroclival meningioma involving upper and middle clivus, Meckel's cave, anterior region of petrosal bone, with deformation of brainstem and cerebellum

Fig. 13 The left presigmoid craniotomy was performed with a combined posterior petrosectomy and subtemporal approach. Posterior semicircular canal (arrowhead) and posterior edge of sigmoid sinus were exposed

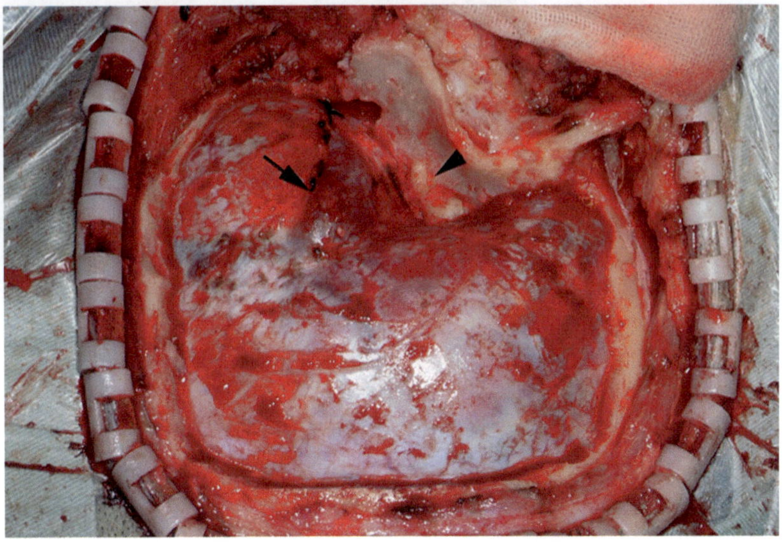

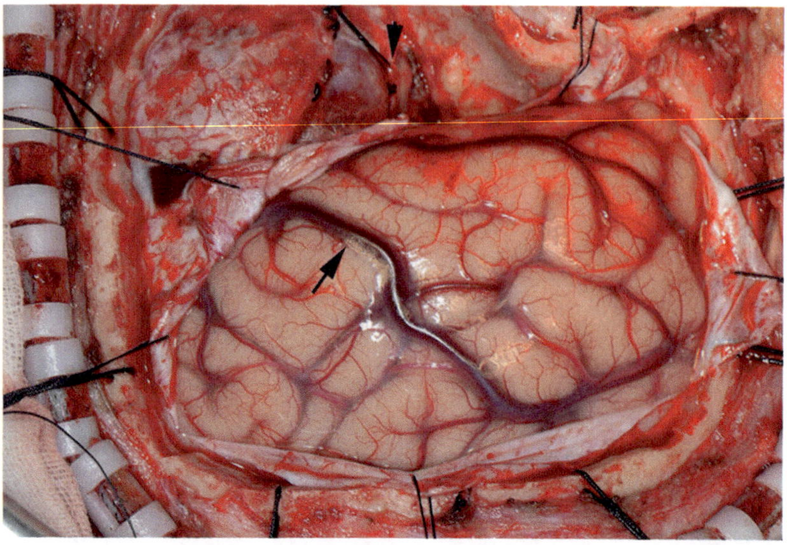

Fig. 14 After dura opening, the vein of Labbe (arrow) was seen on the surface of temporal lobe

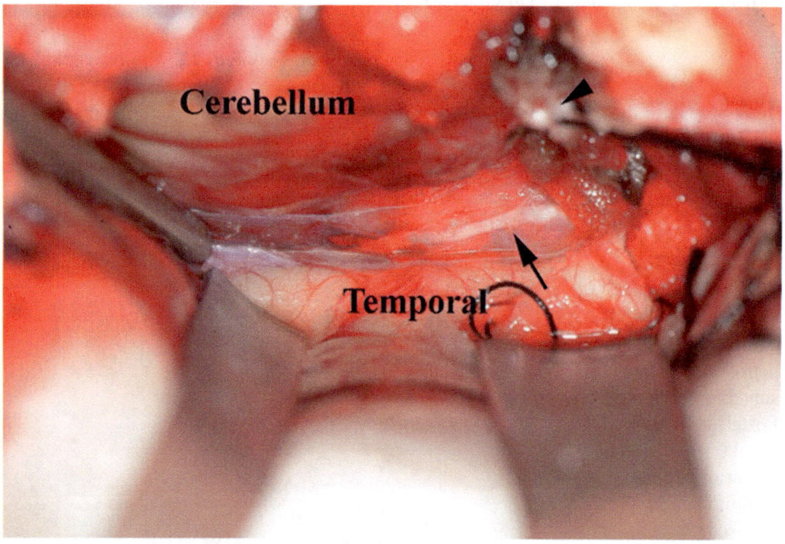

Fig. 15 Basal surface of temporal lobe was elevated followed by cutting cerebellar tentorium (arrowhead), basal cistern was approached with embedded 4th nerve (arrow)

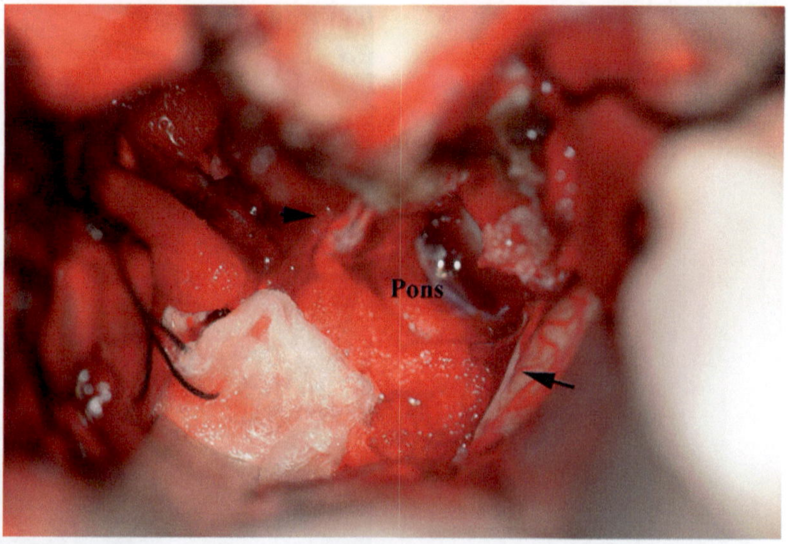

Fig. 16 After tumor resection, trochlear (long arrow) and trigeminal nerve (short arrow) were both seen and preserved

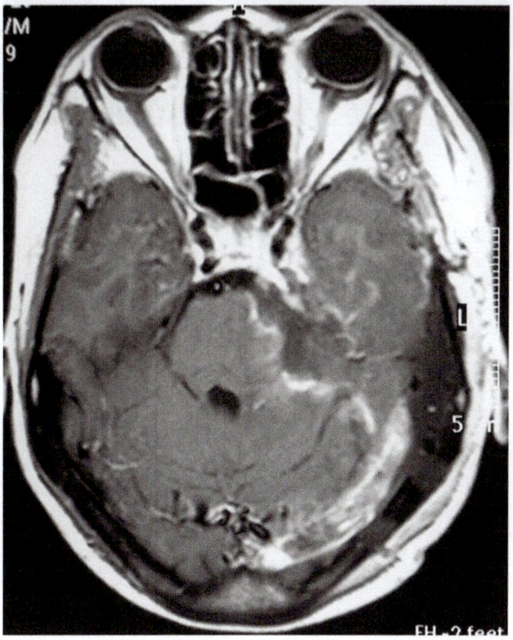

Fig. 17 Post-op axial T1 Gd-enhanced MRI

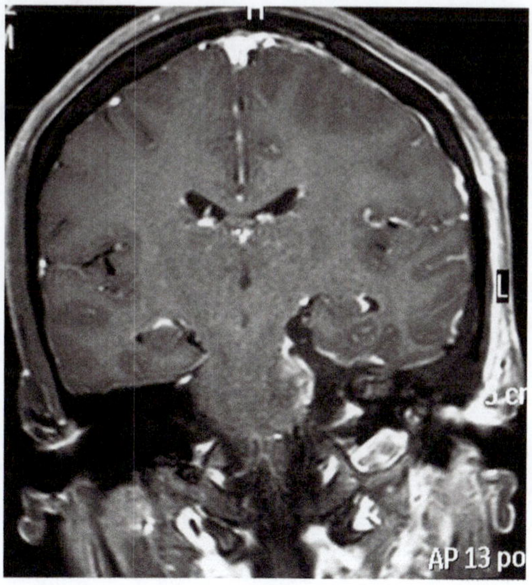

Fig. 18 Post-op coronal T1 Gd-enhanced MRI

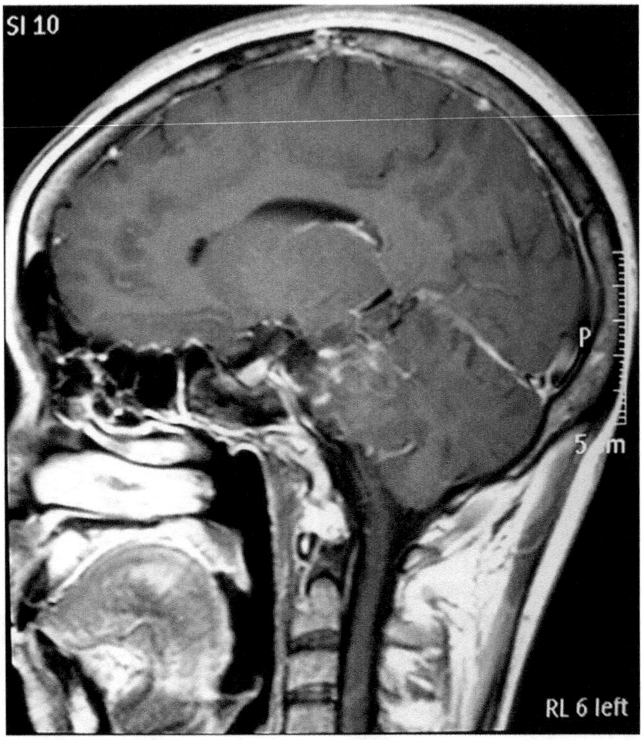

Fig. 19 Post-op sagittal T1 Gd-enhanced MRI. Post-op T1 Gd-enhanced MRI revealed total resection of mass

Anatomy of Cerebellum, 4th Ventricle and Surgical Techniques

Xiang'En Shi, Long Wang, and Hai Qian

1 Anatomy of Cerebellum

Cerebellar cortex has three aspects and each surface has its own approach. The tentorial surface that faced tentorium was reached by supracerebellar infratentorial (Krause) approach. Suboccipital surface, which is located between transverse and sigmoid sinuses was approached by suboccipital craniotomy. The petrosal surface that faced forward to the posterior part of petrosal bone was exposed via retrosigmoid approach. A variety of cerebellar lobules (semilunar lobule, biventral lobule, etc.) were separated by the cortical fissures (suboccipital fissure, postclival fissure, etc.) and massive fissures were naturally created between cerebellum and brainstem (cerebellopontine fissure, cerebellomedullary fissure, etc.).

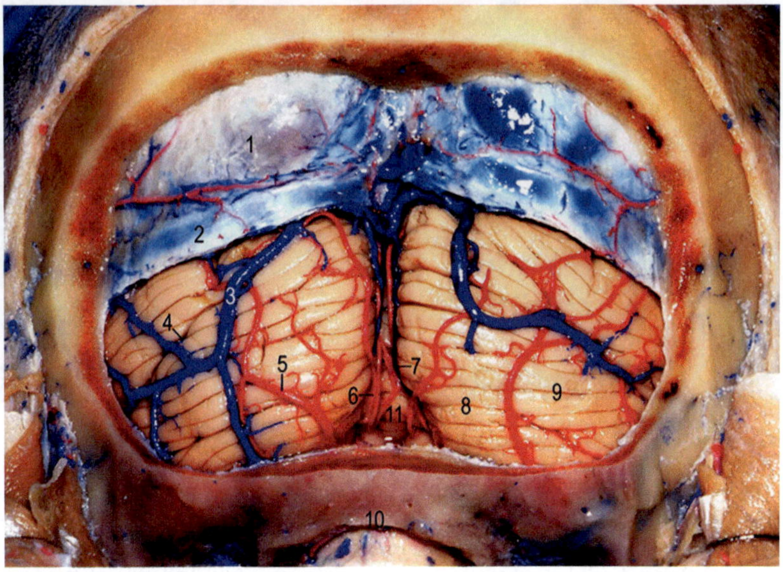

Fig. 1 Posterior view of posterior fossa. (*1*) Occipital dura mater, (*2*) Transverse sinus, (*3*) Inferior hemispheric vein, (*4*) Petrosal/horizontal fissure, (*5*) Lateral trunk of posterior inferior cerebellar artery, (*6*) Medial trunk of posterior inferior cerebellar artery, (*7*) Inferior vermis, (*8*) Medial biventral lobule, (*9*) Lateral biventral lobule, (*10*) Posterior lip of foramen magnum, (*11*) Pyramid

X. Shi (✉) · L. Wang · H. Qian
Neurological Surgery, SanBo Brain Hospital, Capital Medical University, Beijing, China
e-mail: shixen@ccmu.edu.cn; wanglong@ccmu.edu.cn; qianh@ccmu.edu.cn

X. Shi et al. (eds.), *Microneurosurgical Anatomy and Surgical Technique*,
https://doi.org/10.1007/978-981-19-8273-6_6

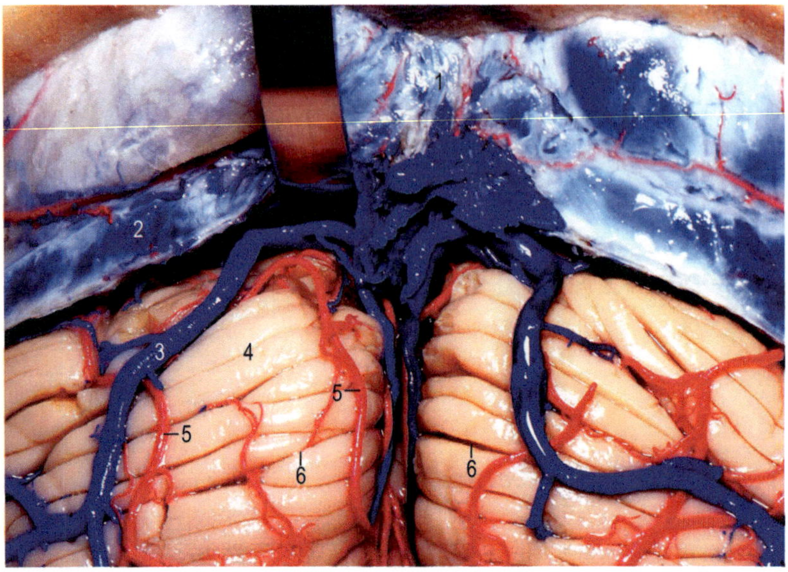

Fig. 2 Posterior view of superior portion of occipital cerebellum. (*1*) Confluence of sinuses, (*2*) Transverse sinus, (*3*) Inferior hemispheric vein, (*4*) Inferior semilunar lobule, (*5*) Cortical segment of superior cerebellar artery, (*6*) Suboccipital fissure

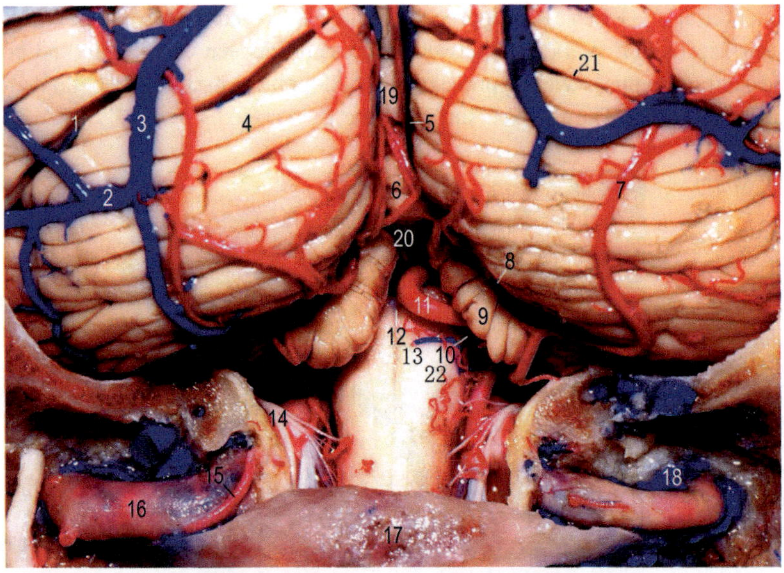

Fig. 3 Posterior view of inferior portion of occipital cerebellum and craniocervical region. (*1*) Petrosal/horizontal fissure, (*2*) Inferior hemispheric vein of inferior semilunar lobule, (*3*) Inferior hemispheric vein of occipital fissure, (*4*) Inferior semilunar lobule, (*5*) Posterior cerebellar incisura, (*6*) Pyramid, (*7*) Lateral trunk of posterior inferior cerebellar artery, (*8*) Tonsillobiventral fissure, (*9*) Cerebellar Tonsil, (*10*) Tonsillomedullary fissure, (*11*) Caudal loop of posterior inferior cerebellar artery, (*12*) Obex, (*13*) Fasciculus gracilis, (*14*) Lower cranial nerves, (*15*) Posterior meningeal artery, (*16*) Transverse portion of atlantoaxial segment of vertebral artery, (*17*) Posterior arch of atlas, (*18*) Vertebral venous plexus, (*19*) Tuber, (*20*) Vallecula of the cerebellum, (*21*) Suboccipital fissure, (*22*) Fasciculus cuneatus

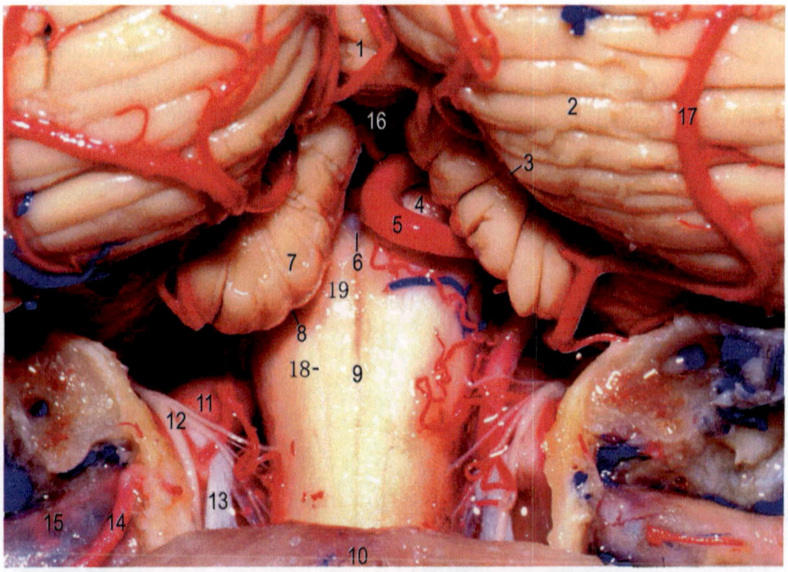

Fig. 4 Enlarged view of craniocervical region. (*1*) Pyramid, (*2*) Biventral lobule, (*3*) Tonsillobiventral fissure, (*4*) Cuneate tubercle, (*5*) Caudal loop of posterior inferior cerebellar artery, (*6*) Obex, (*7*) Cerebellar Tonsil, (*8*) Tonsillomedullary fissure, (*9*) Posterior median sulcus, (*10*) Posterior arch of atlas, (*11*) Intracranial segment of vertebral artery, (*12*) Lower cranial nerves, (*13*) Dentate ligament, (*14*) Posterior meningeal artery, (*15*) Oblique portion of atlantoaxial segment of vertebral artery, (*16*) Vallecula of the cerebellum, (*17*) Lateral trunk of posterior inferior cerebellar artery, (*18*) Posterior intermediate sulcus, (*19*) Gracile tubercle

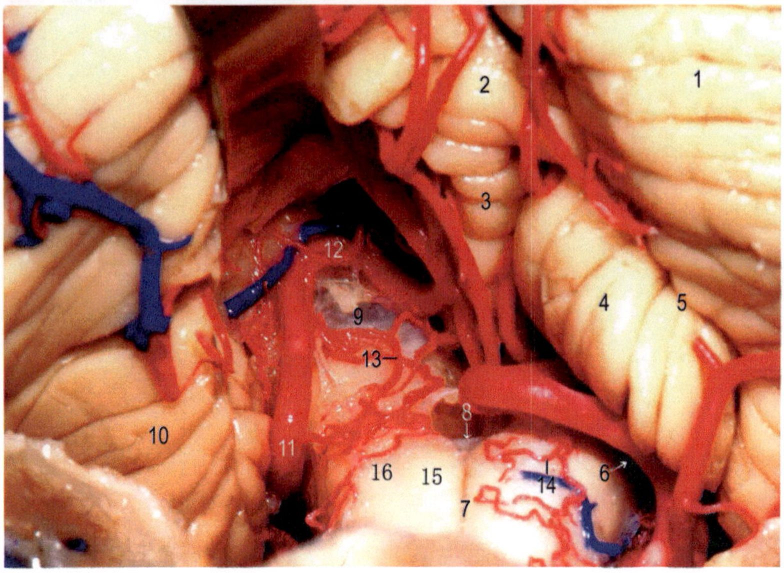

Fig. 5 Exposure of dorsal brainstem with left tonsil resection. (*1*) Right biventral lobule, (*2*) Pyramid, (*3*) Uvula of vermis, (*4*) Cerebellar Tonsil, (*5*) Right tonsillobiventral fissure, (*6*) Right tonsillomedullary fissure, (*7*) Posterior median sulcus, (*8*) Obex, (*9*) Tela choroidea of fourth ventricle, (*10*) Left biventral lobule, (*11*) Caudal loop of posterior inferior cerebellar artery, (*12*) Cranial loop of posterior inferior cerebellar artery, (*13*) Choroidal arteries of posterior inferior cerebellar artery, (*14*) Posterior spinal artery, (*15*) Gracile tubercle, (*16*) Cuneate tubercle

2 Anatomy of 4th Ventricle

The 4th ventricle is a tent-shaped space between brainstem ventrally and cerebellum dorsally. This cavity communicated 3rd ventricle through cerebral aqueduct at rostral and caudally connected to the cisterna magna via foramen of Magendie. Laterally, the CSF of 4th ventricle flew through the foramen of Luschka and then join cerebellopontine cistern.

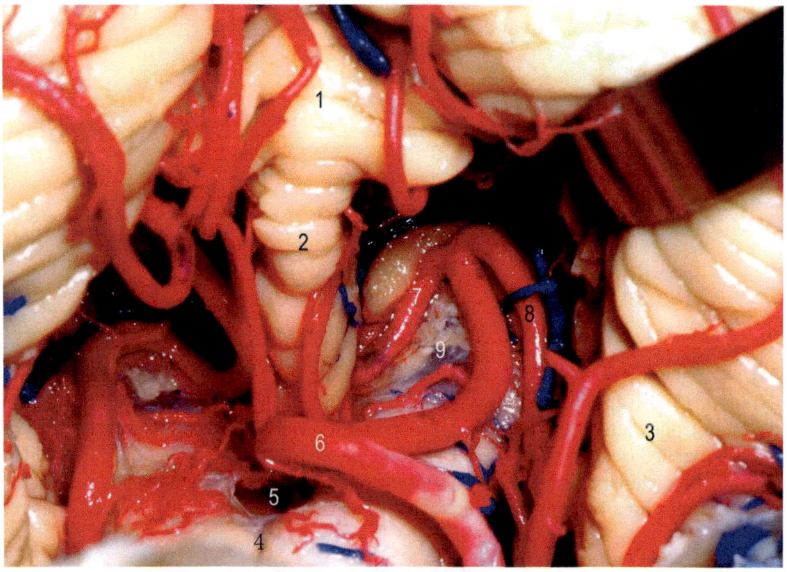

Fig. 6 Surgical trajectory with superolateral retraction of tonsil via tonsillomedullary fissure. (*1*) Cerebellar pyramid, (*2*) Uvula of vermis, (*3*) Biventral lobule, (*4*) Posterior median sulcus, (*5*) Foramen of Magendie, (*6*) Tonsillomedullary segment of posterior inferior cerebellar artery, (*7*) Choroidal arteries of posterior inferior cerebellar artery, (*8*) Telovelotonsillar segment of posterior inferior cerebellar artery, (*9*) Tela choroidea

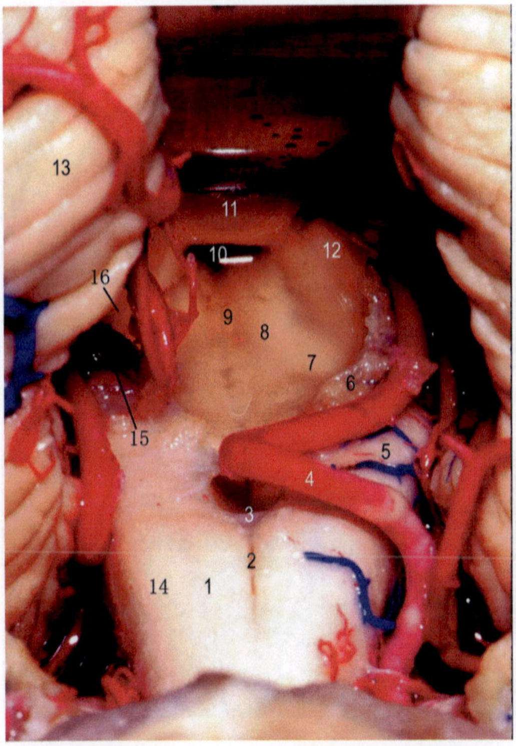

Fig. 7 Exposure of dorsal medulla oblongata and 4th ventricle. (*1*) Gracile tubercle, (*2*) Posterior medial sulcus, (*3*) Obex, (*4*) Posterior inferior cerebellar artery, (*5*) Inferior cerebellar peduncle, (*6*) Choroid plexus, (*7*) Vestibular area, (*8*) Medial eminence, (*9*) Median sulcus of medulla oblongata, (*10*) Cerebral aqueduct of midbrain, (*11*) Superior medullary velum, (*12*) Superior cerebellar peduncle, (*13*) Biventral lobule, (*14*) Cuneate tubercle, (*15*) Lateral recess, (*16*) Dentate tubercle

Case 6.1 Cerebellopontine Angle Meningioma

Clinical presentation:

A 55-year-old female presented with a 2-year history of intermittent left tinnitus and a 1-month history of left facial numbness.

Neurological exam:

Left facial hypoesthesia

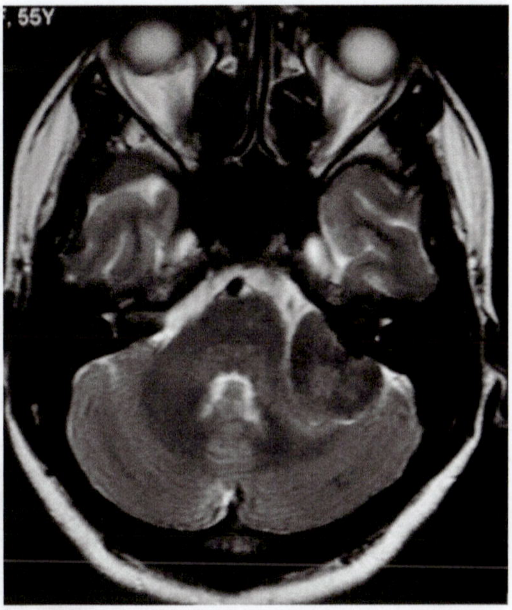

Fig. 8 Pre-op axial T2-weighted MRI

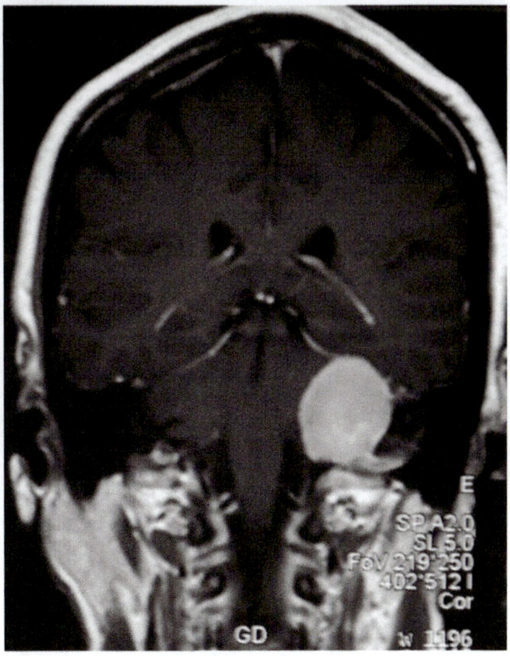

Fig. 9 Pre-op coronal Gd-enhanced T1-weighted MRI

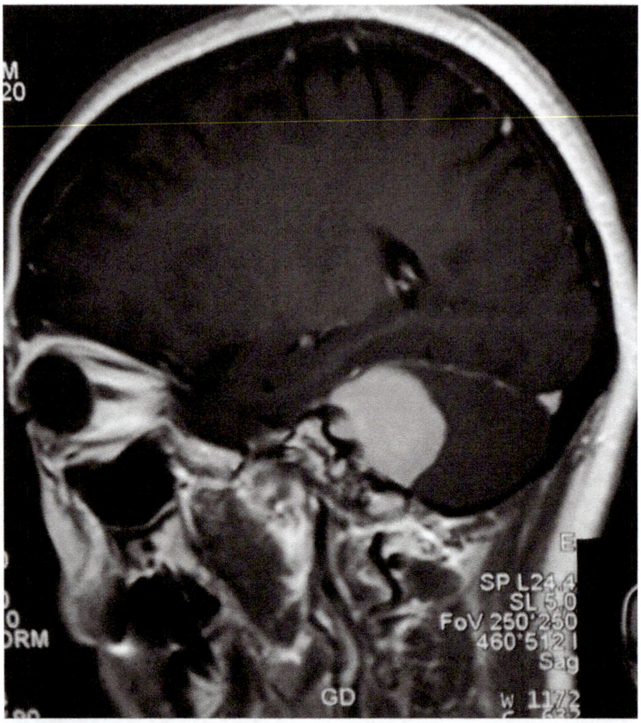

Fig. 10 Pre-op sagittal Gd-enhanced T1-weighted MRI. Pre-op MR images revealed a giant, well-marginated cerebellopontine angle mass with deformation of pontine and cerebellum. The lesion was homogeneously enhanced with "dura tail" sign at posterior part of petrosal bone

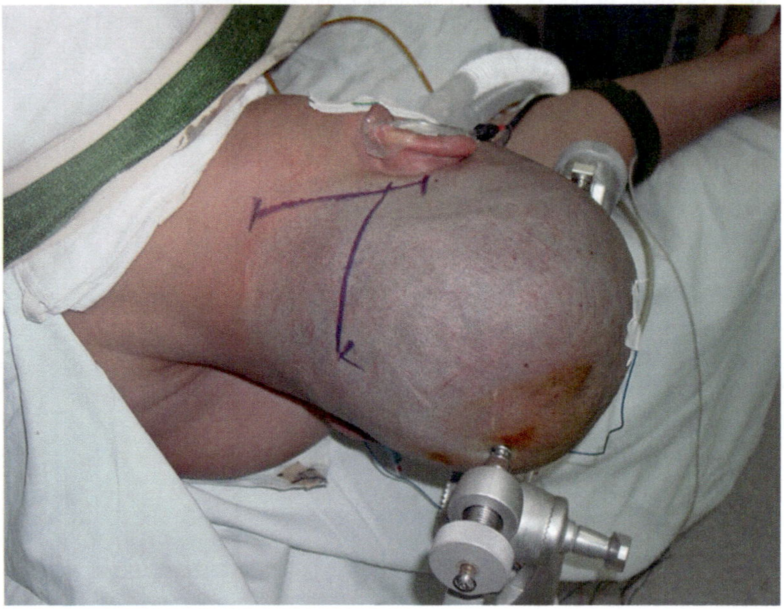

Fig. 11 The patient was placed in a lateral decubitus position and a "reverse 7" incision was made for this giant mass

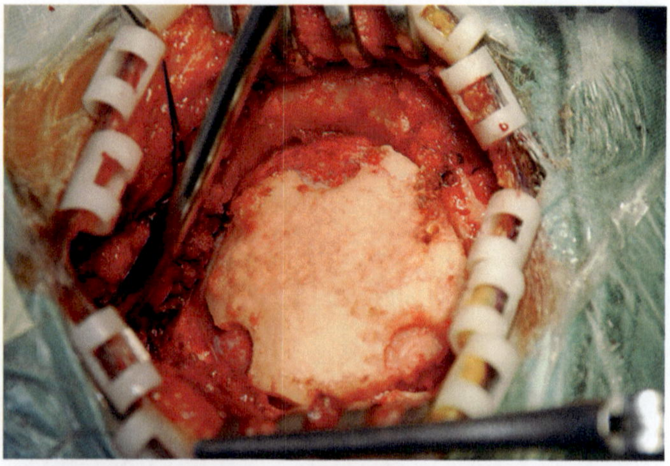

Fig. 12 The retrosigmoid craniotomy was carried out to expose transverse–sigmoid junction and posterior edge of sigmoid sinus

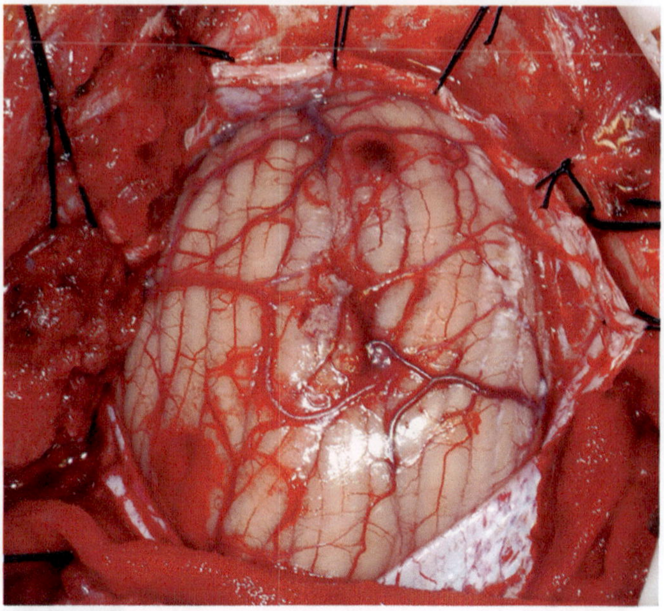

Fig. 13 Dura was opened and occipital aspect of cerebellum was exposed

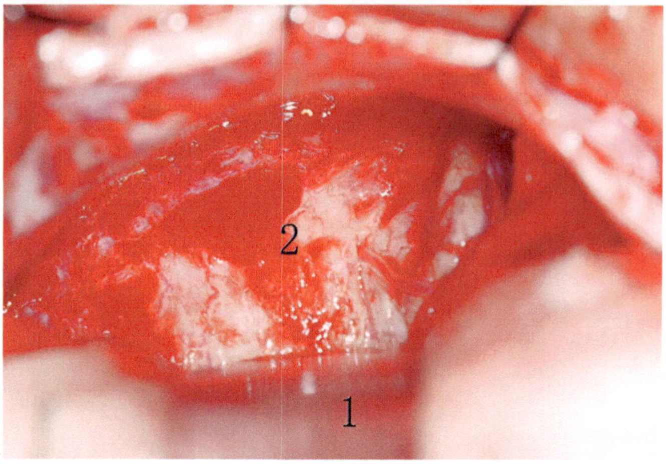

Fig. 14 The tumor was seen following medial retraction of cerebellum; (*1*) retraction spatula; (*2*) meningioma

Fig. 15 Cranial nerves were preserved after total resection of tumor, (*1*) Trigeminal nerve, (*2*) facial–vestibulocochlear complex, (*3*) abducens nerve, (*4*) lower cranial nerves, and (*5*) left cerebellar hemisphere

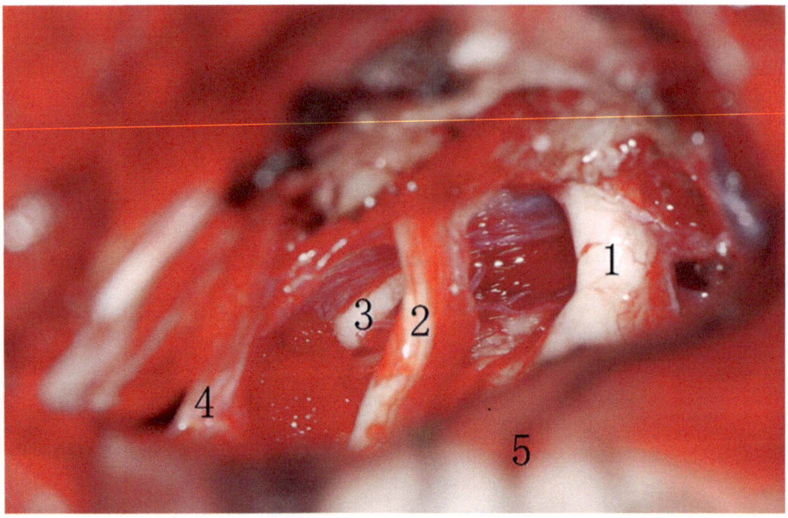

Case 6.2 Cerebellar Pilocytic Astrocytoma

Clinical presentation:

A 5-year-old girl presented with a 3-month history of headache with visual deficits.

Neurological exam:

Normal

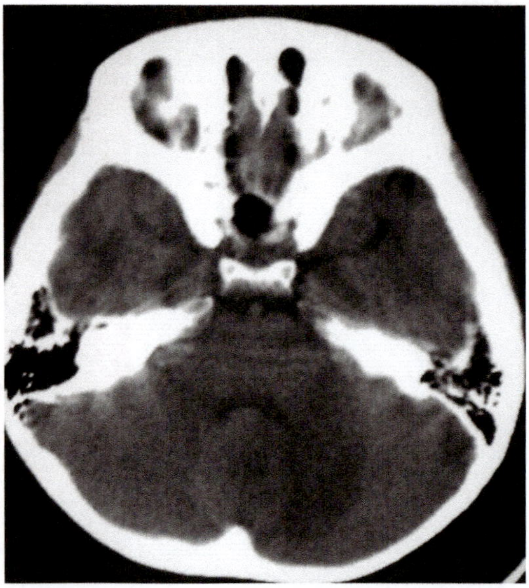

Fig. 16 Axial CT revealed a solid hypodense cerebellar mass

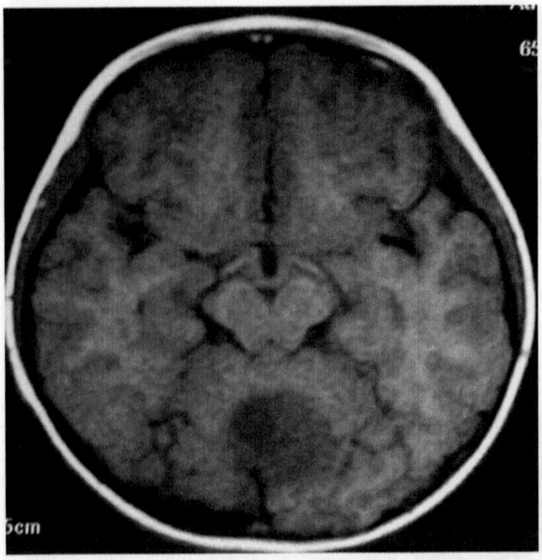

Fig. 17 Pre-op axial T1-weighted MRI

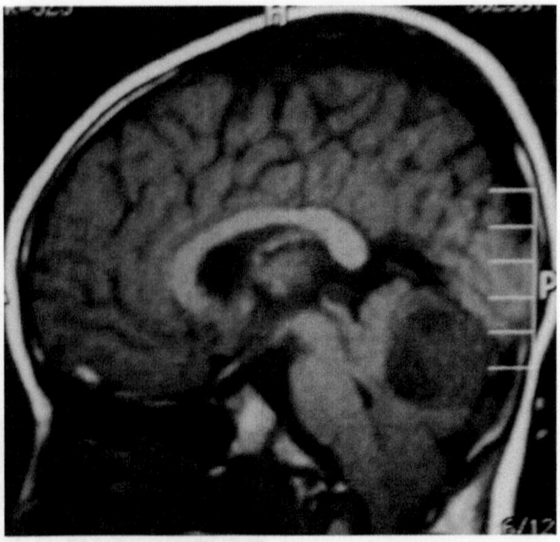

Fig. 18 Pre-op sagittal T1-weighted MRI

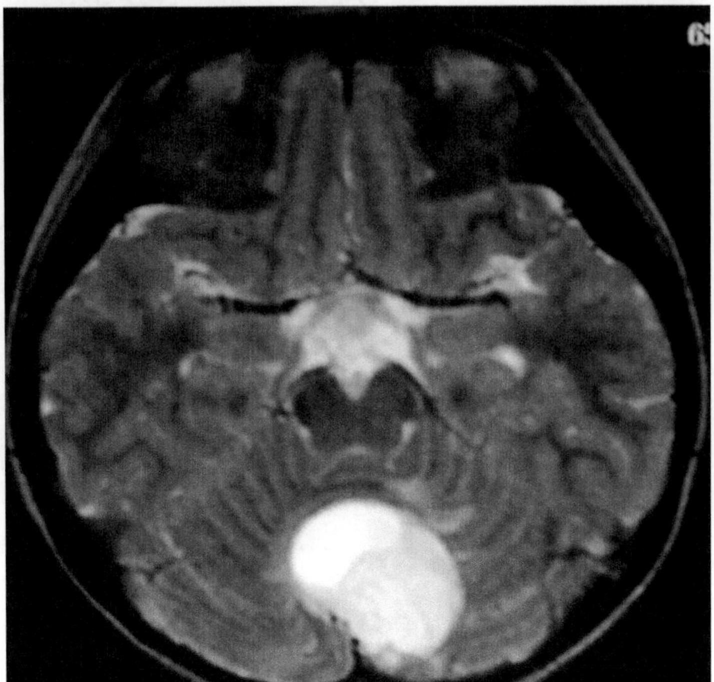

Fig. 19 Pre-op axial T2-weighted MRI. Pre-op MRI illustrated a well-marginated round lesion

Fig. 20 The patient was placed at a park bench position and midline suboccipital incision was created

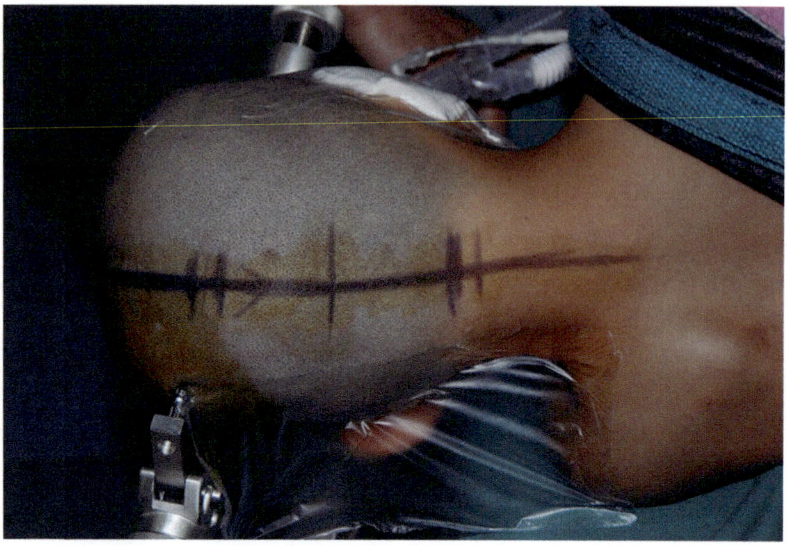

Fig. 21 Two burr holes were made over the transverse sinus on either side of midline. Then bone flap was inscribed reaching the foramen magnum

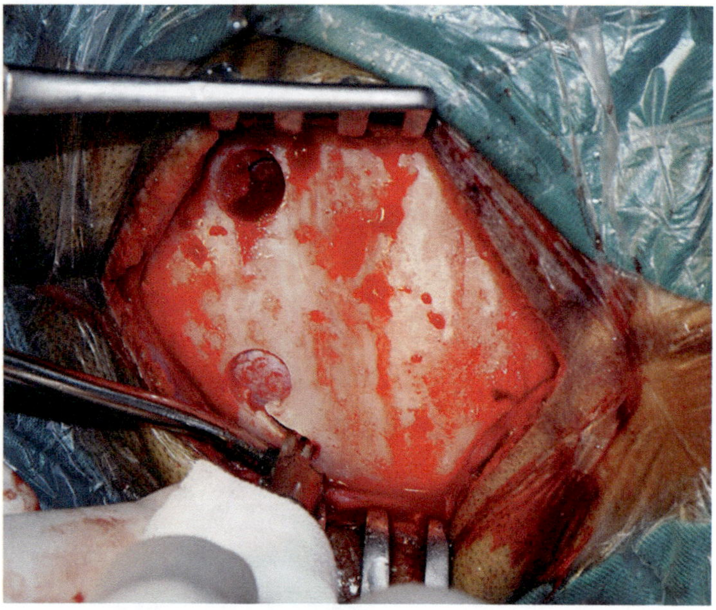

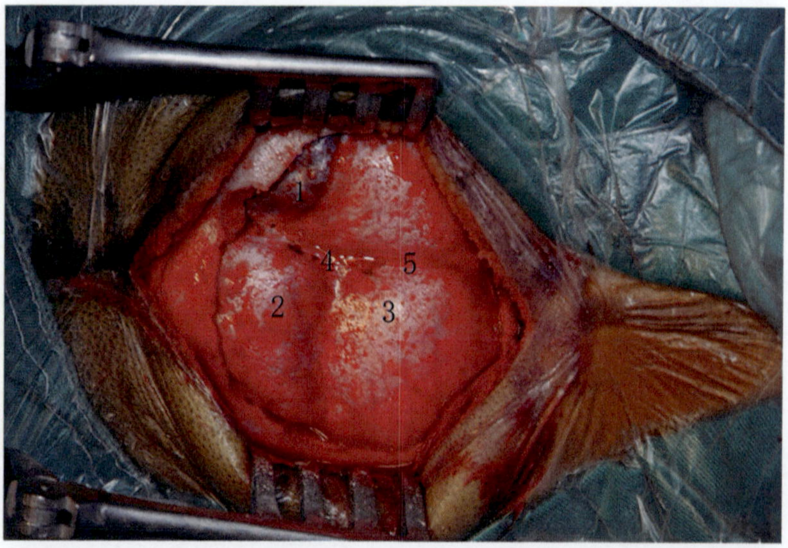

Fig. 22 The dominance of right transverse sinus was seen after bone flap removal. (*1*) Transverse sinus, (*2*) left occipital lobe, (*3*) left occipital surface of cerebellum, (*4*) torcular herophili, (*5*) occipital sinus

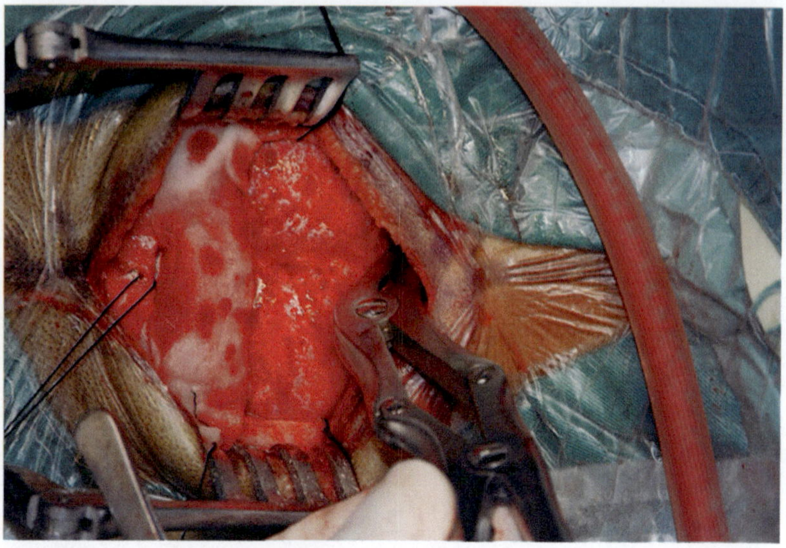

Fig. 23 The bone flap was extended dorsally to open the posterior wall of foramen magnum

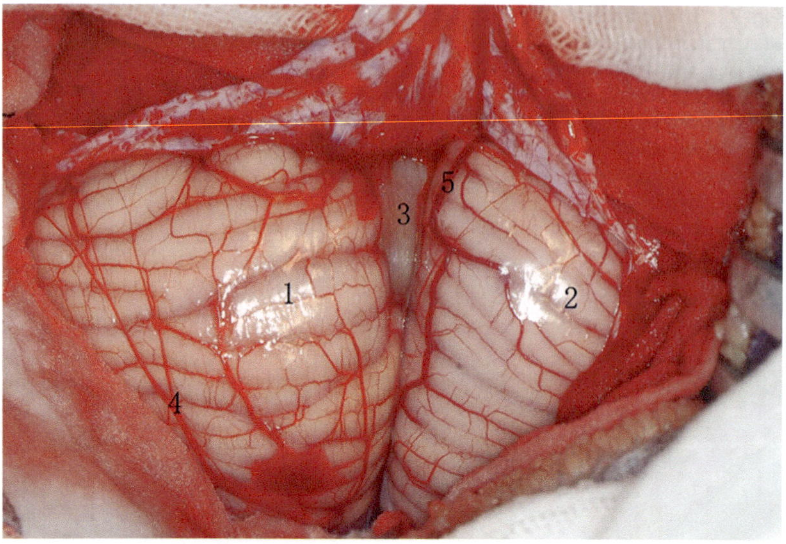

Fig. 24 Dura was opened to expose intradural structures. (*1*) Left cerebellar hemisphere, (*2*) Right cerebellar hemisphere, (*3*) tumor, (*4*) cortical branches of posterior inferior cerebellar artery, and (*5*) cortical branches of superior cerebellar artery

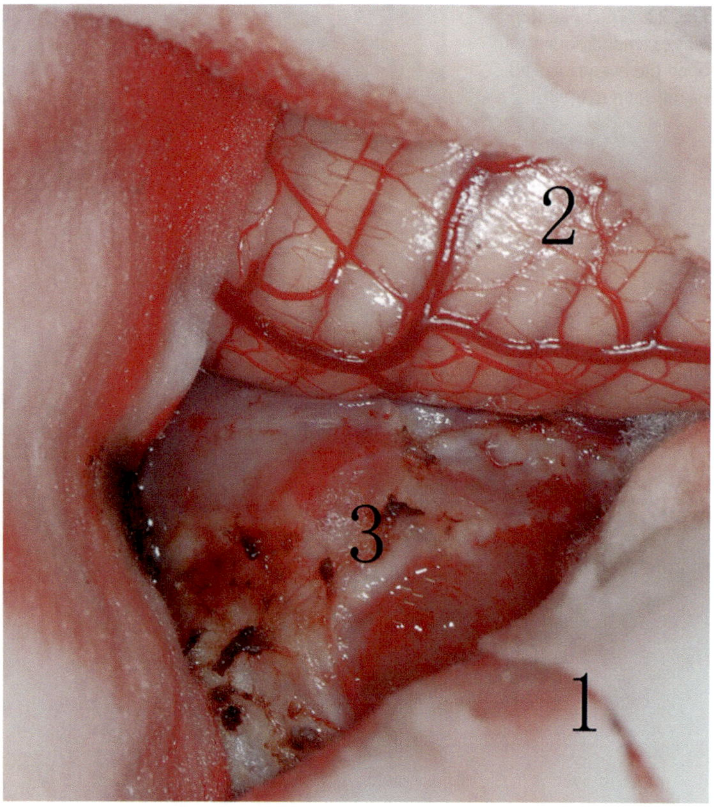

Fig. 25 The mass arise from vermis was further exposed with bilateral retraction of cerebellum. (*1*) Left cerebellar hemisphere, (*2*) right cerebellar hemisphere, and (*3*) tumor

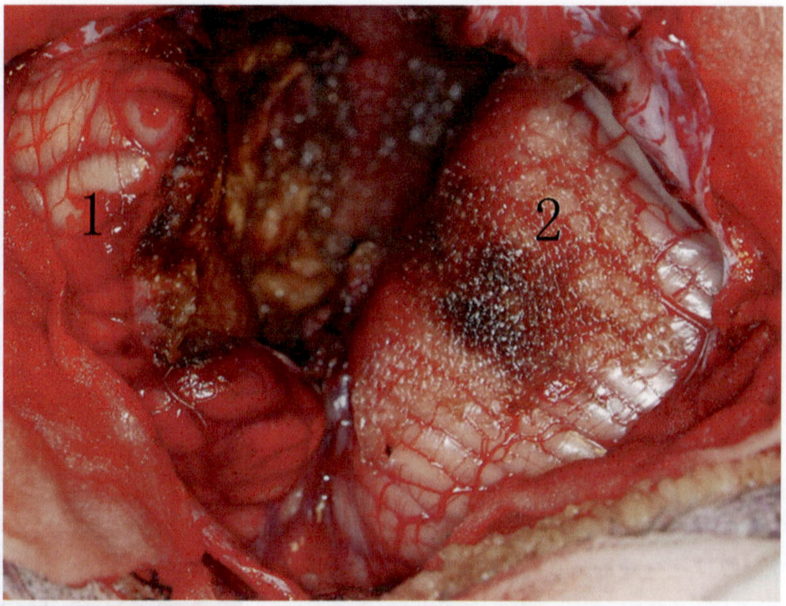

Fig. 26 The exposed surgical field revealed total resection of vermis mass; (*1*) Left cerebellar hemisphere and (*2*) right cerebellar hemisphere

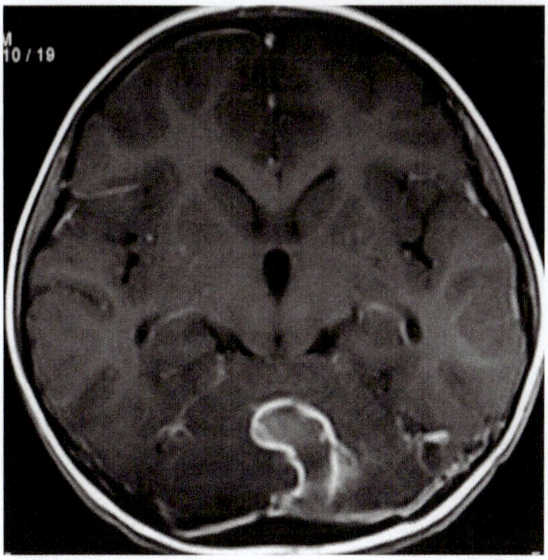

Fig. 27 Post-op axial T1-weighted MRI

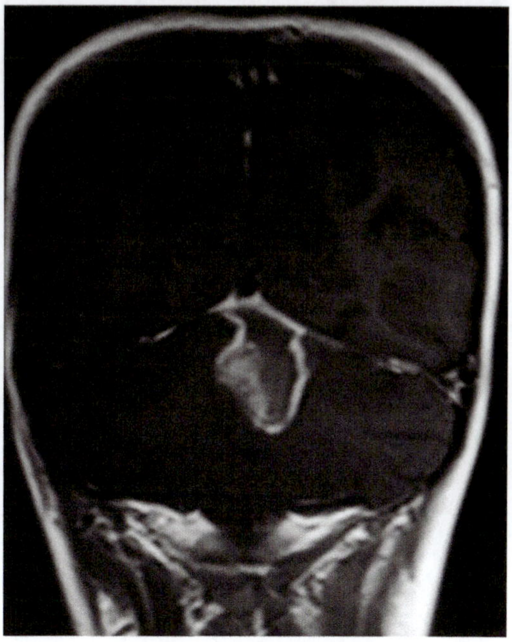

Fig. 28 Post-op coronal T1-weighted MRI. Post-op MRI confirmed disappearance of tumor

Case 6.3 Fusiform Vertebral Artery Aneurysm

Clinical presentation:

A 40-year-old man presented with a 1-month history of severe headache and vomiting.

Neurological exam:

Lethargy, staggering gait, finger-to-nose test (+), Romberg's sign(+)

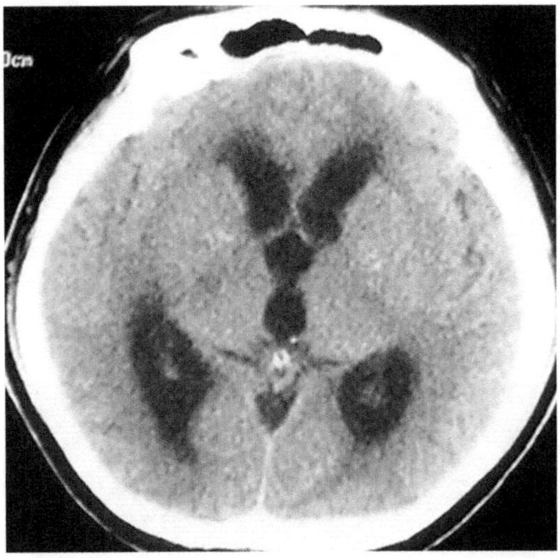

Fig. 29 Pre-op axial CT revealed enlargement of ventricular system

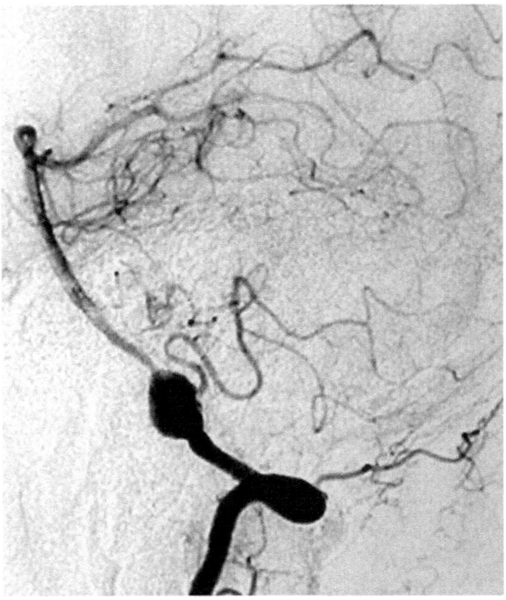

Fig. 30 Pre-op lateral view of right vertebrobasilar artery angiography

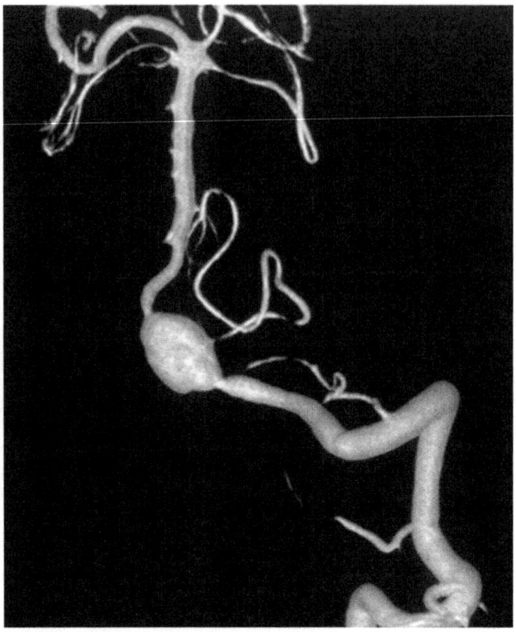

Fig. 31 Pre-op right 3D vertebrobasilar artery angiography. Pre-op DSA showing a right vertebral trunk fusiform aneurysm with involvement of posterior inferior cerebellar artery

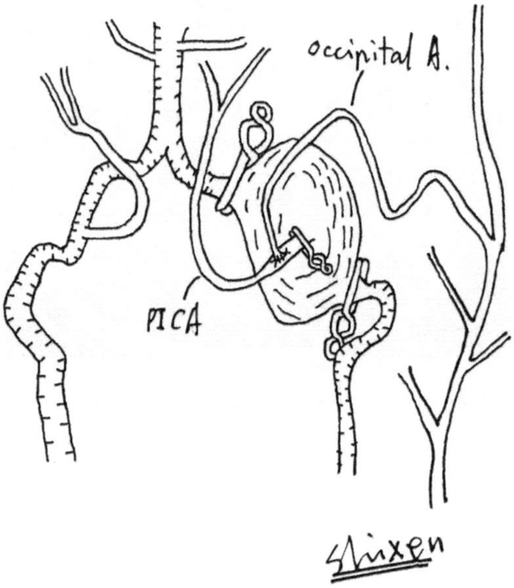

Fig. 32 The medical illustration described surgical strategy. The fusiform V4 aneurysm was totally trapped followed by occipital artery-to-posterior inferior cerebellar artery bypass

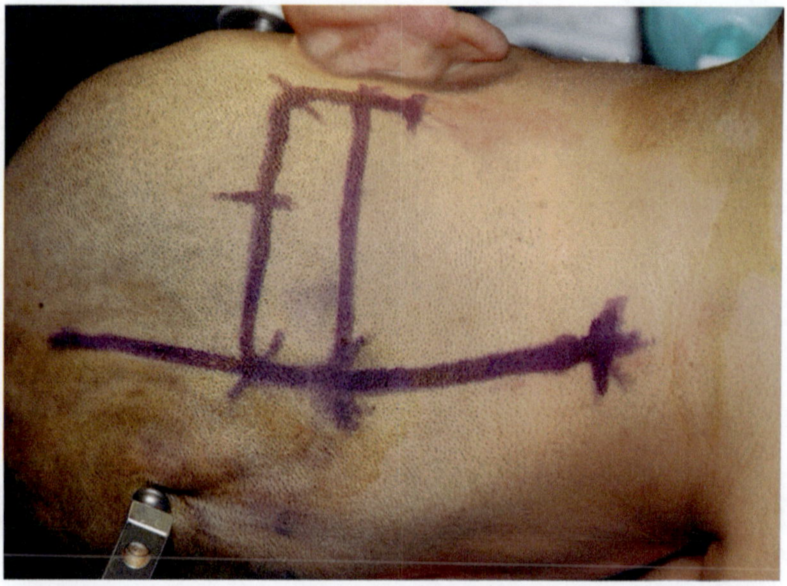

Fig. 33 Left park bench position was utilized with a standard "hockey stick" incision to expose vertebral trunk located at ventral brainstem

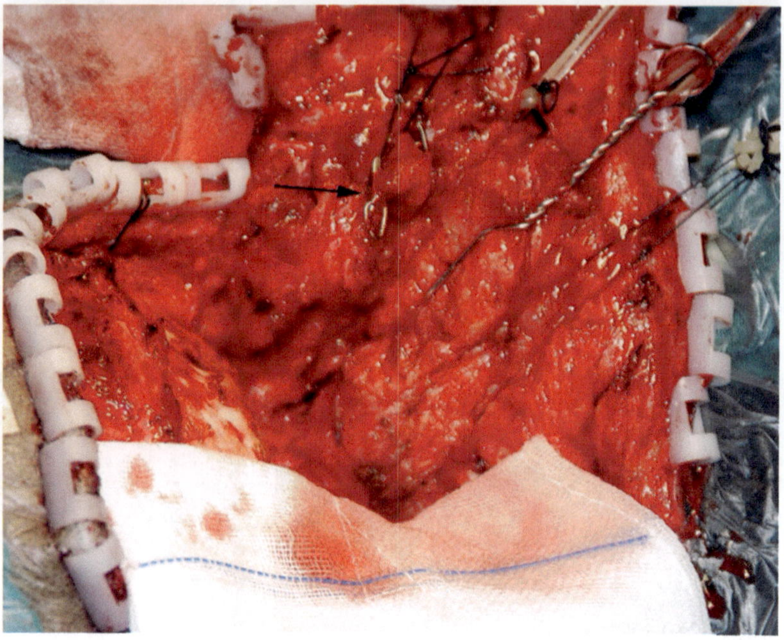

Fig. 34 Occipital artery (arrow) was identified and harvested by a standard outside-in fashion

Fig. 35 The adjunctive tissue and adventitia were cleared during arterial dissection. Pressure distention technique was also carried out to prevent vasospasm before anastomosis

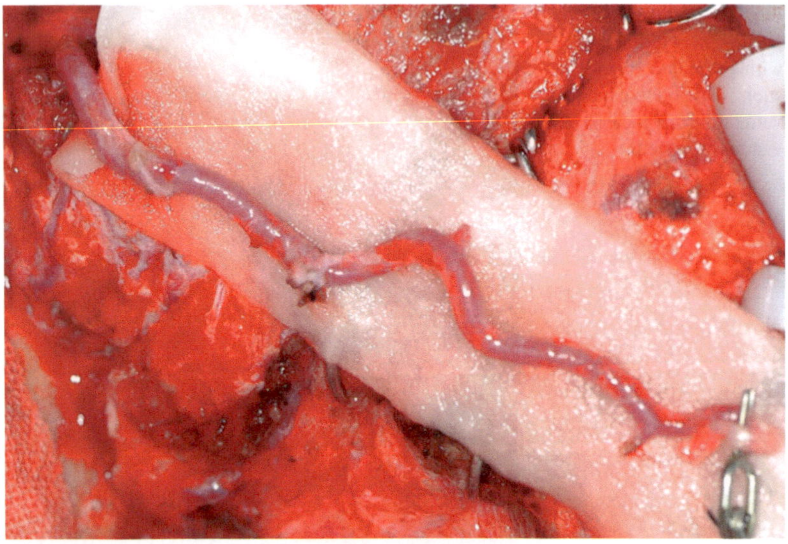

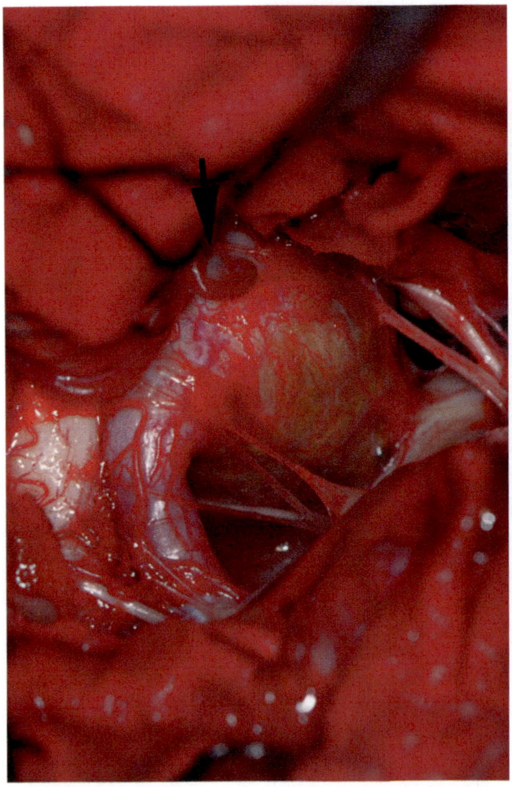

Fig. 36 The fusiform, thrombosed V4 aneurysm was seen with rostromedial retraction of cerebellum. Posterior inferior cerebellar artery (arrow) arising from dome of lesion

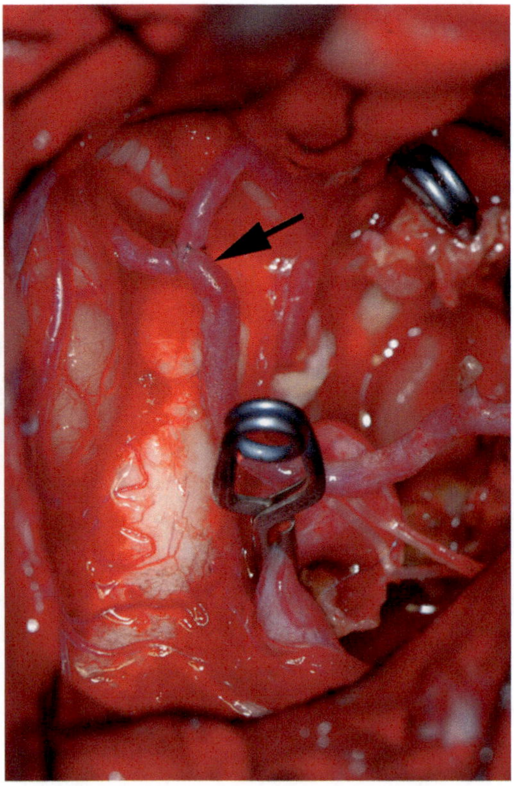

Fig. 37 End-to-side anastomosis was accomplished by connecting occipital artery and caudal loop. Then the aneurysm was trapped and partially removed to decompress the replacement of medulla

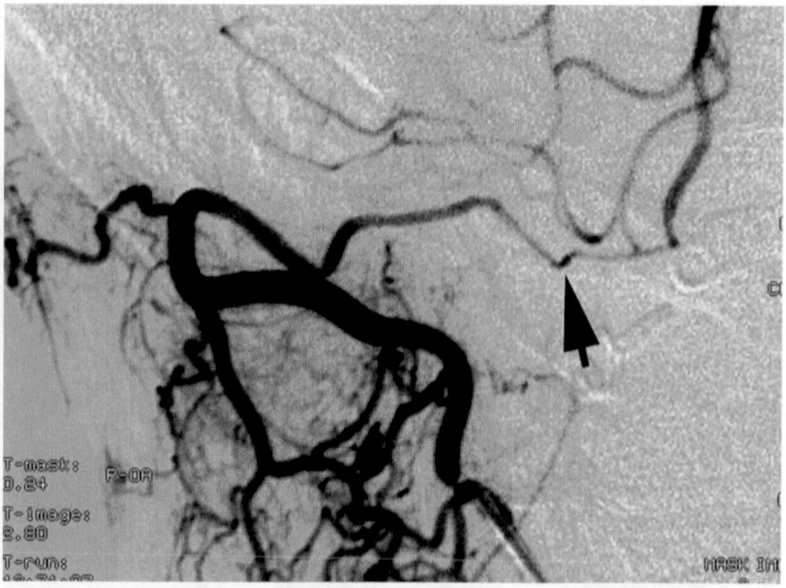

Fig. 38 Post-op oblique view of right external carotid artery angiography showing patent bypass (arrow) status

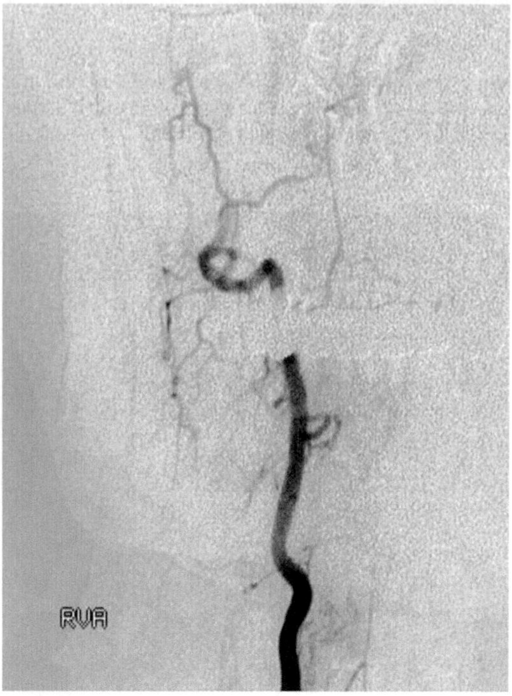

Fig. 39 Post-op oblique view of right vertebral artery angiography revealed disappearance of fusiform aneurysm

Case 6.4 Recurrent Medulla Oblongata Hemangioblastoma

Clinical presentation:

A 27-year-old gentleman presented with a 1-month history of dizziness and 10-day history of dysphagia and hoarse voice.

Neurological exam:

Dysphagia, vocal cord paralysis, deviation of uvula, loss of gag reflex

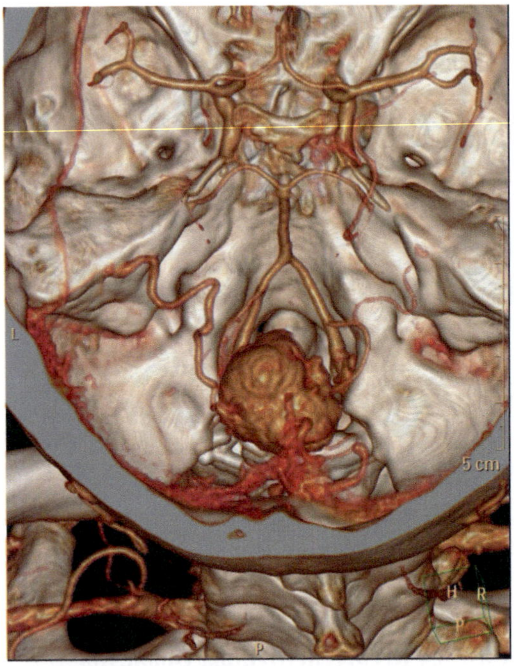

Fig. 40 Pre-op superior view of 3D CTA

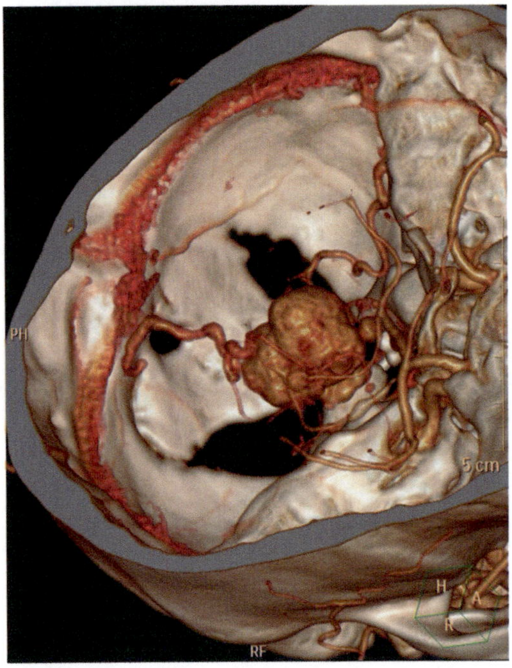

Fig. 41 Pre-op right lateral view of 3D CTA

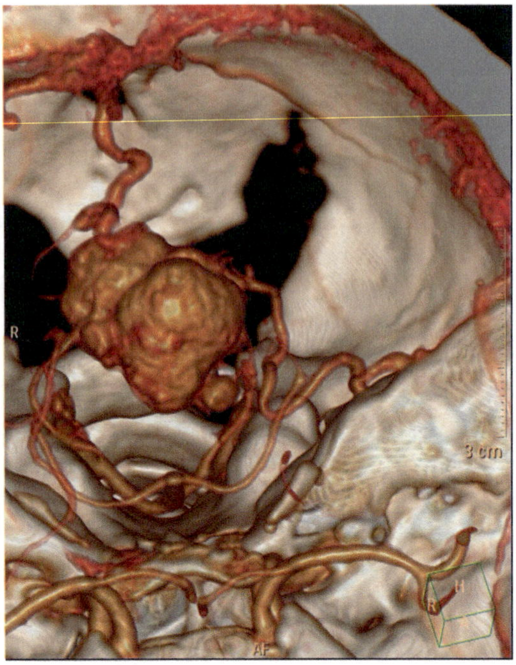

Fig. 42 Pre-op left lateral view of 3D CTA. Pre-op CTA showing the extensive vascularity of a large solid lesion of posterior fossa that was fed by bilateral PICA and drained into straight/superior petrosal sinus

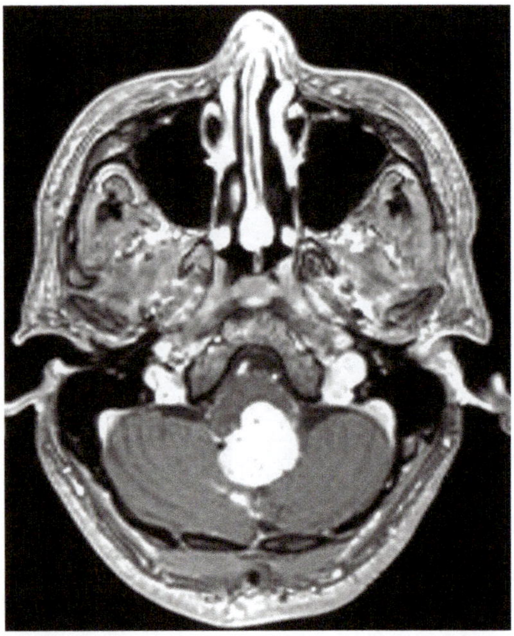

Fig. 43 Pre-op axial Gd-enhanced T1-weighted MRI

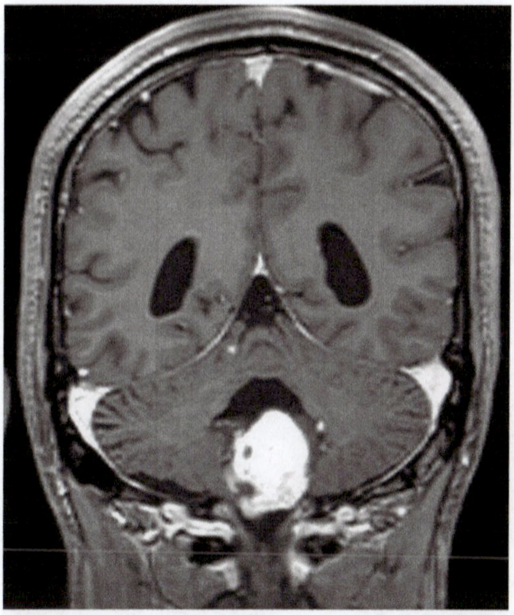

Fig. 44 Pre-op coronal Gd-enhanced T1-weighted MRI

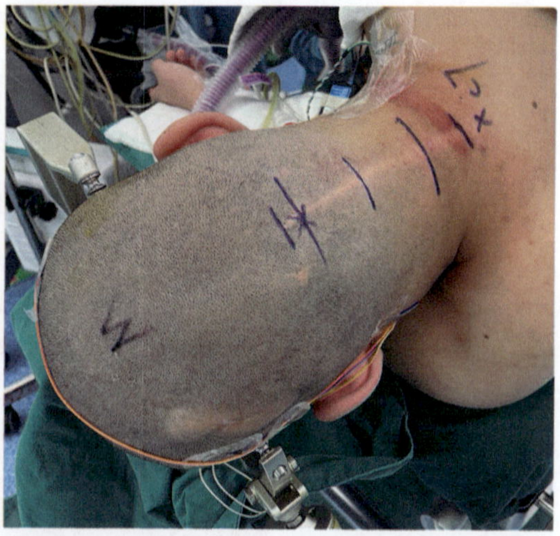

Fig. 46 Patient was placed in a park bench position and suboccipital midline incision was marked

Fig. 45 Pre-op sagittal Gd-enhanced T1-weighted MRI. Pre-op Gd-enhanced MRI revealed a purely solid mass at dorsal medullary oblongata with avidly enhancing presentation

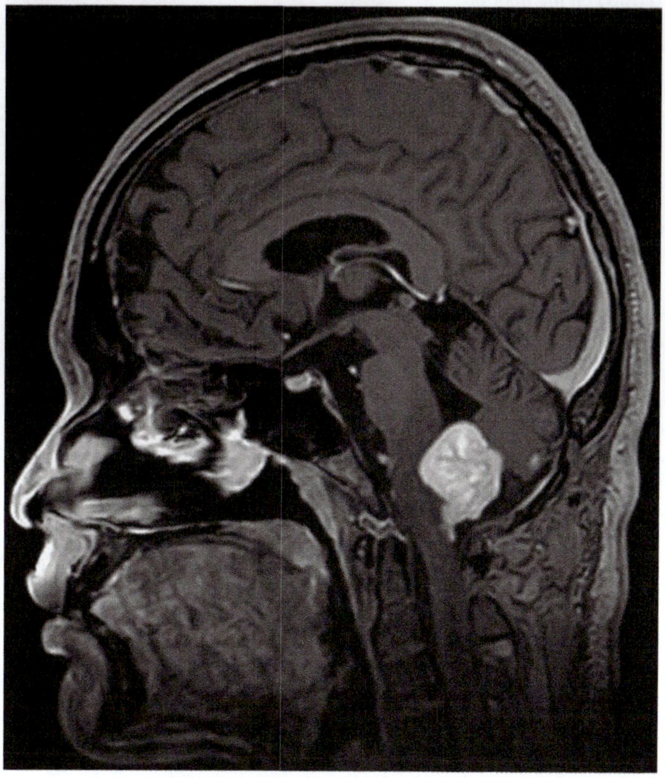

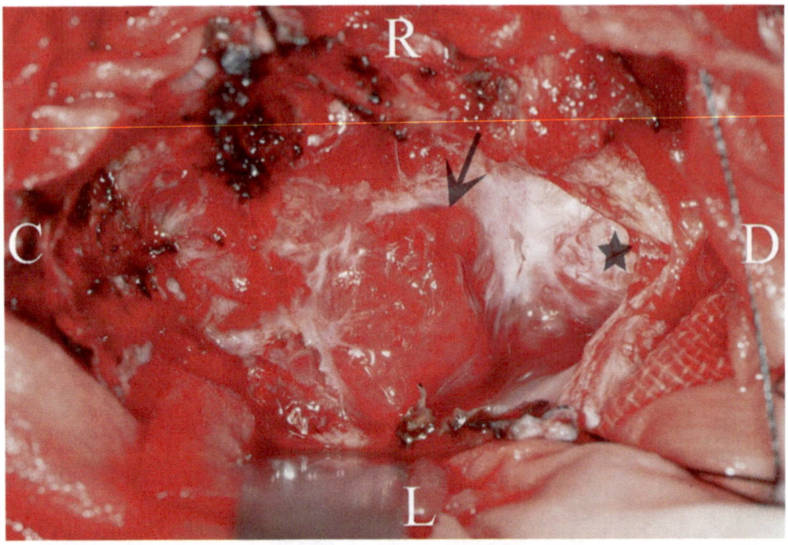

Fig. 47 The hypervascularized nodule (arrow) and upper cervical spine (asterisk) were unveiled with medial retraction of cerebellar tonsil. (C, cranial; D, dorsal; L, left; R, right)

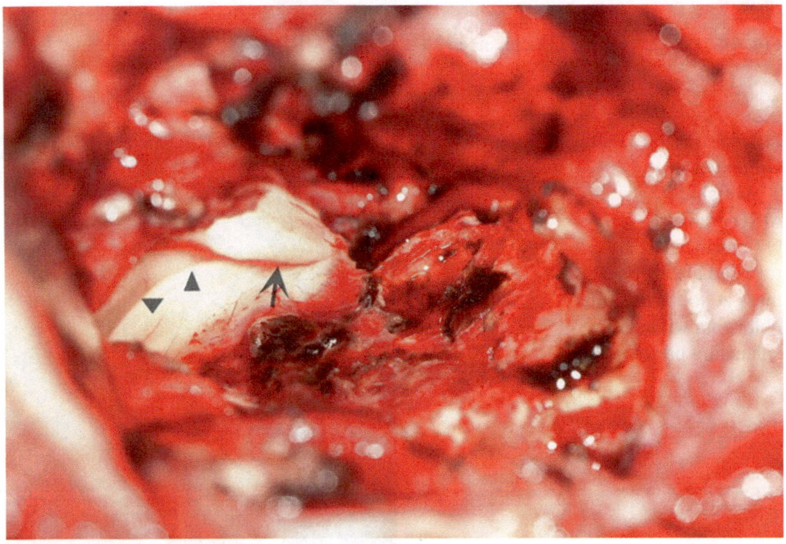

Fig. 48 After mass resection, stria medullaris (arrowhead) and median sulcus (arrow) on the floor of 4th ventricle were identified

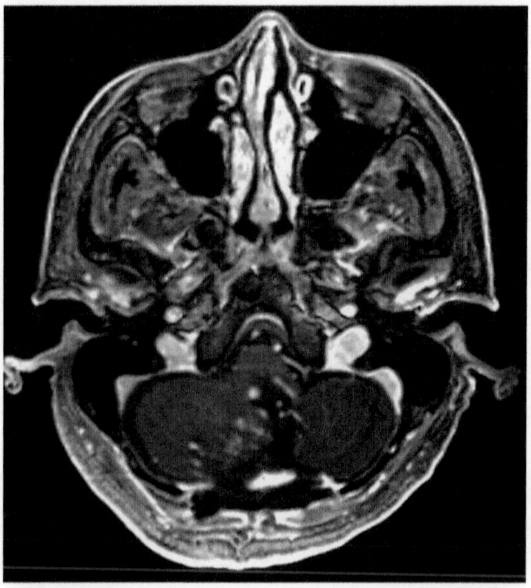

Fig. 49 Post-op axial Gd-enhanced T1-weighted MRI

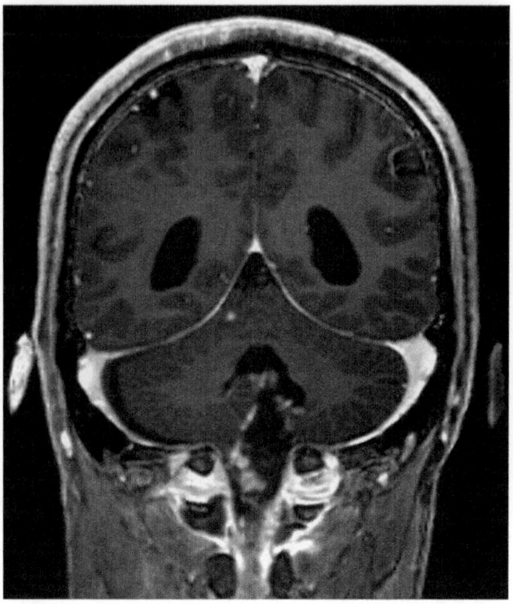

Fig. 50 Post-op coronal Gd-enhanced T1-weighted MRI

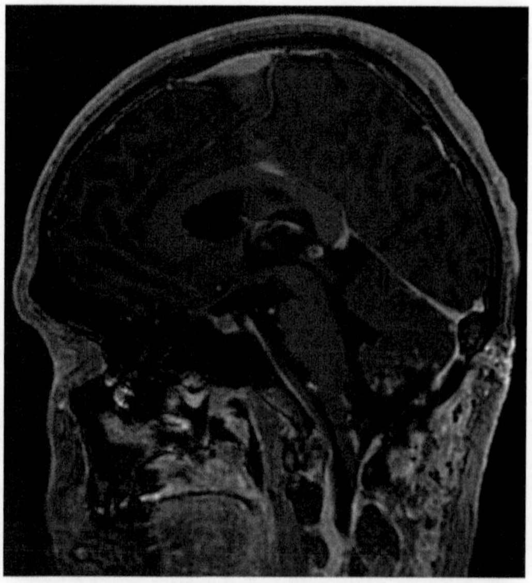

Fig. 51 Post-op sagittal Gd-enhanced T1-weighted MRI

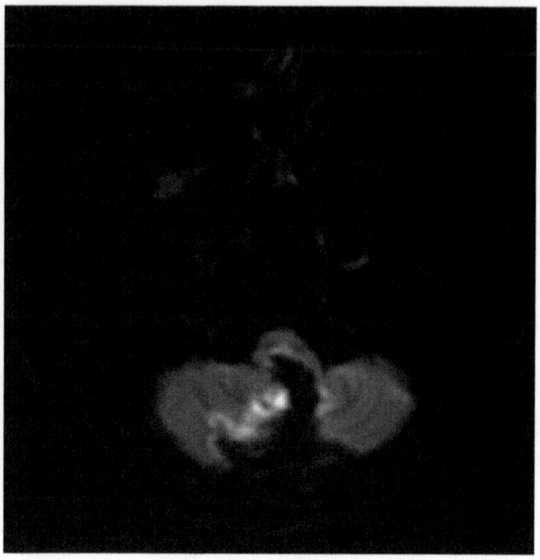

Fig. 52 Post-op DWI. Post-op MRI showed the mass was aggressively removed without adjacent complications

Anatomy of Craniocervical Junction, Cervical Region and Surgical Techniques

Xiang'En Shi, Long Wang, and Hai Qian

1 Anatomy of Craniocervical Junction

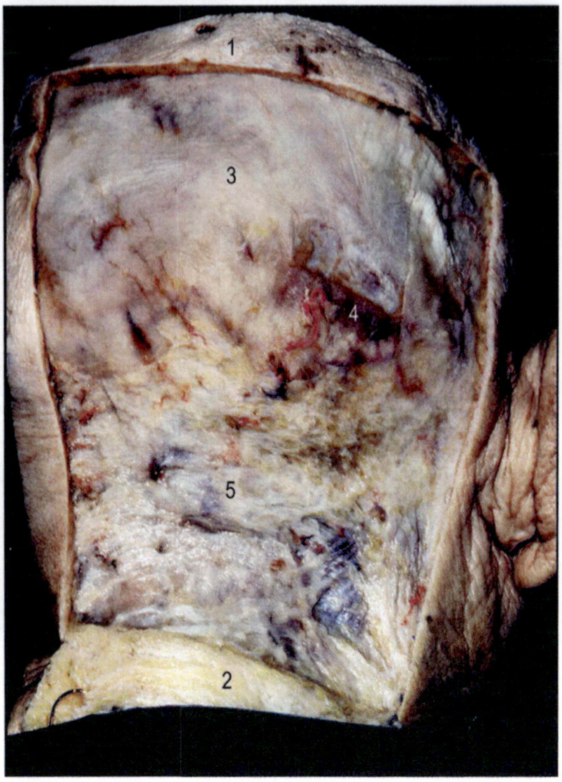

Fig. 1 Subcutaneous layer of craniocervical junction. (*1*) Scalp, (*2*) Subcutaneous tissue, (*3*) Galea aponeurotica, (*4*) Occipitalis, (*5*) Trapezius

X. Shi (✉) · L. Wang · H. Qian
Neurological Surgery, SanBo Brain Hospital, Capital Medical University, Beijing, China
e-mail: shixen@ccmu.edu.cn; wanglong@ccmu.edu.cn; qianh@ccmu.edu.cn

X. Shi et al. (eds.), *Microneurosurgical Anatomy and Surgical Technique*,
https://doi.org/10.1007/978-981-19-8273-6_7

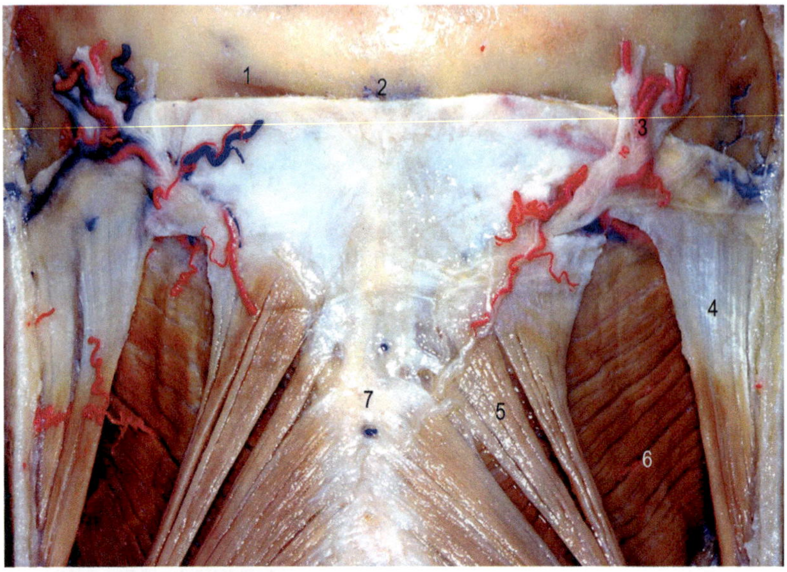

Fig. 2 Superficial and 1st deep muscular layer of craniocervical junction. (*1*) Superior nuchal line, (*2*) External occipital protuberance, (*3*) Occipital artery, (*4*) Sternocleidomastoid muscle, (*5*) Trapezius muscle, (*6*) Splenius capitis muscle, (*7*) Nuchal ligament

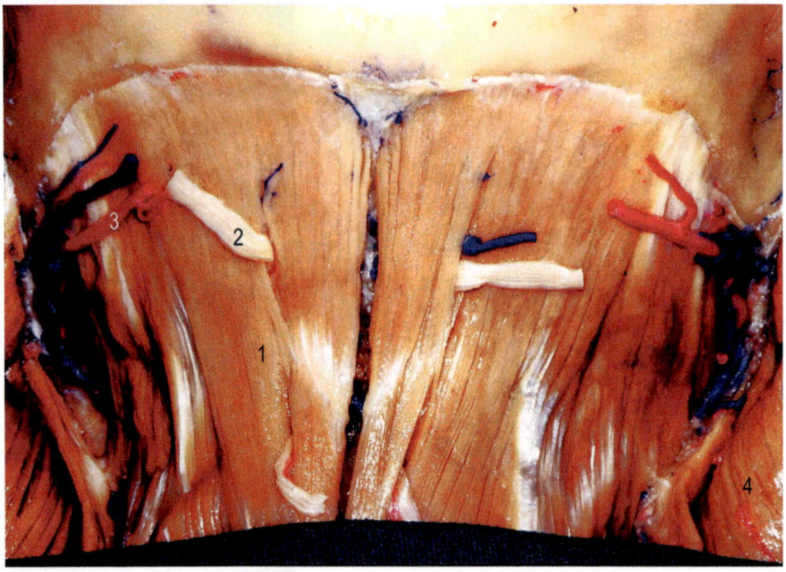

Fig. 3 2nd deep muscular layer of craniocervical junction. (*1*) Semispinalis capitis muscle, (*2*) Greater occipital nerve, (*3*) Suboccipital segment of occipital artery, (*4*) Splenius capitis muscle

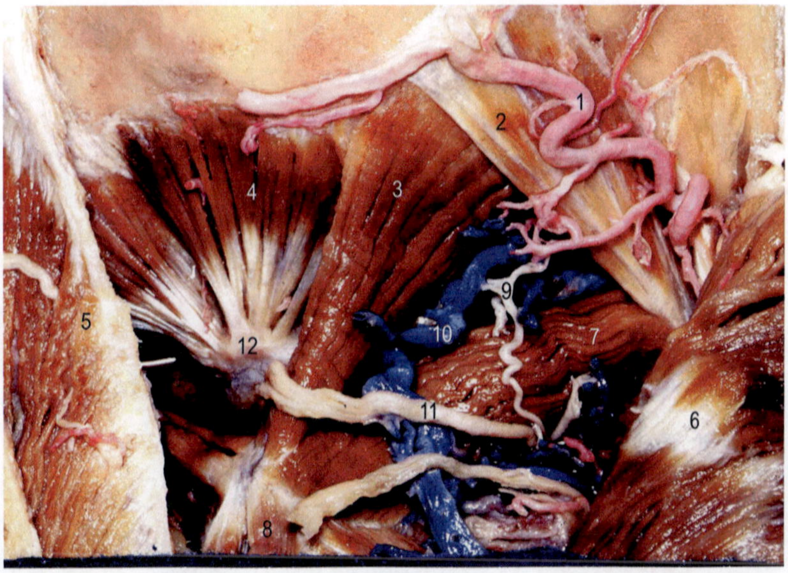

Fig. 4 3rd deep muscular layer of craniocervical junction. (*1*) Suboccipital segment of occipital artery, (*2*) Obliquus capitis superior, (*3*) Rectus capitis posterior major, (*4*) Rectus capitis posterior minor, (*5*) Trapezius muscle, (*6*) Semispinalis capitis muscle, (*7*) Obliquus capitis inferior, (*8*) Interspinales cervicis, (*9*) Dorsal ramus branch of C_1 cervical spinal nerve, (*10*) Suboccipital venous plexus, (*11*) Dorsal ramus branch of C_2 cervical spinal nerve, (*12*) Posterior arch of atlas

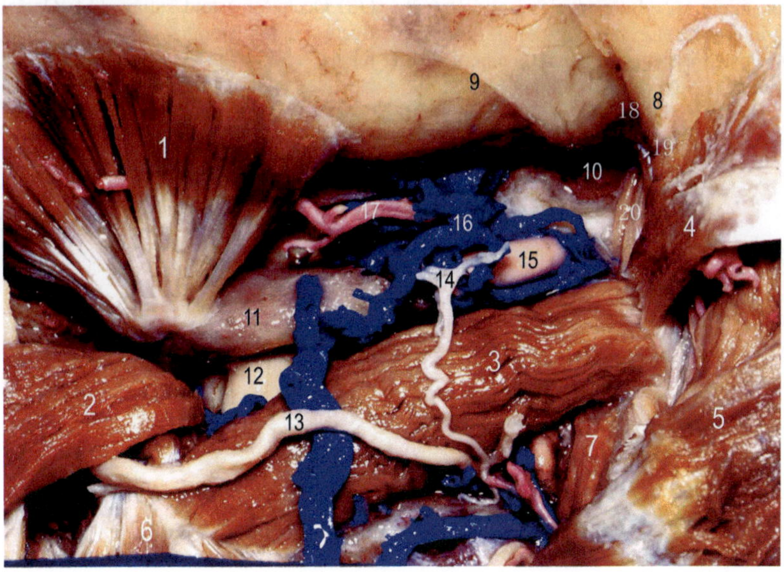

Fig. 5 Overview of suboccipital triangle. (*1*) Rectus capitis posterior minor, (*2*) Rectus capitis posterior major, (*3*) Obliquus capitis inferior posterior oblique muscle, (*4*) Obliquus capitis superior posterior oblique muscle, (*5*) Semispinalis capitis muscle, (*6*) Interspinales cervicis, (*7*) Levator muscle of scapula, (*8*) Occipitomastoid suture, (*9*) Inferior nuchal line, (*10*) Occipital condyle, (*11*) Posterior arch of atlas, (*12*) Posterior cervical spine meninges, (*13*) Dorsal ramus root of C2, (*14*) Dorsal ramus Root of C1, (*15*) Vertebral artery, (*16*) Suboccipital venous plexus, (*17*) Artery of Salmon, (*18*) Occipital groove, (*19*) Posterior belly of digastric muscle, (*20*) Lateral Rectus capitis

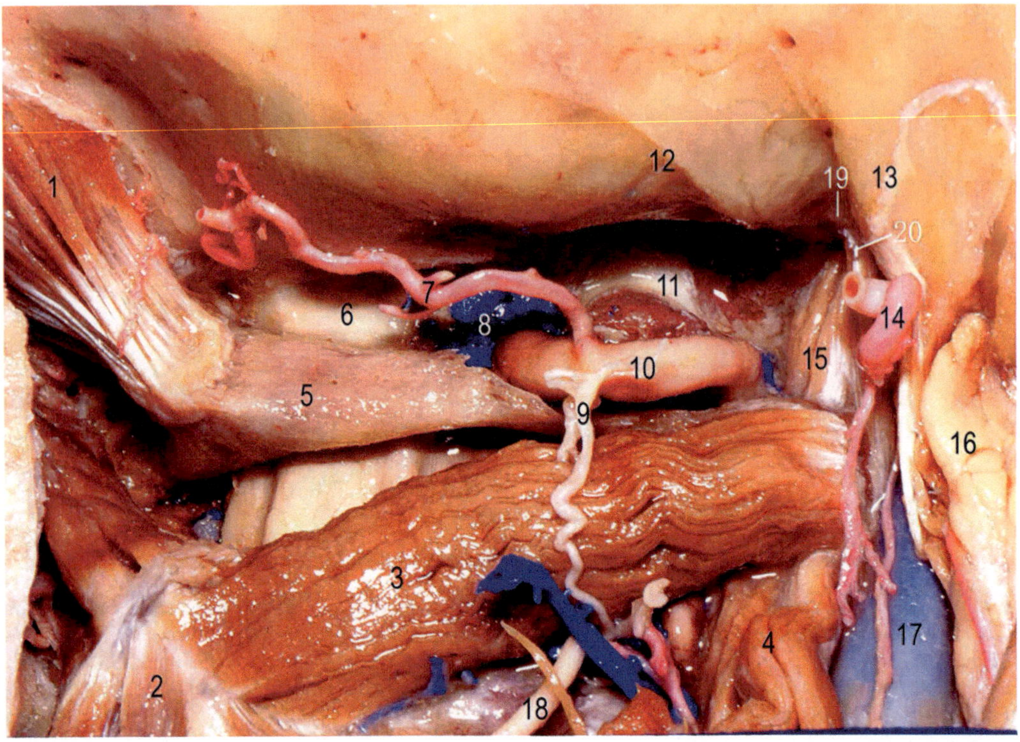

Fig. 6 Enlarged view of suboccipital triangle. (*1*) Rectus capitis posterior minor, (*2*) Interspinales cervicis, (*3*) Obliquus capitis inferior, (*4*) Levator scapula, (*5*) Posterior arch of atlas, (*6*) Posterior atlantooccipital membrane, (*7*) Artery of Salmon, (*8*) Suboccipital venous plexus, (*9*) Dorsal ramus Roots of C1 spinal nerve, (*10*) Vertebral artery, (*11*) Occipital condyle, (*12*) Inferior nuchal line, (*13*) Occipitomastoid suture, (*14*) Occipital artery, (*15*) Rectus capitis lateralis, (*16*) Parotid gland, (*17*) Internal jugular vein, (*18*) Dorsal ramus of C2 spinal nerve, (*19*) Occipital groove, (*20*) Posterior belly of digastric muscle

2 Anatomy of Cervical Region

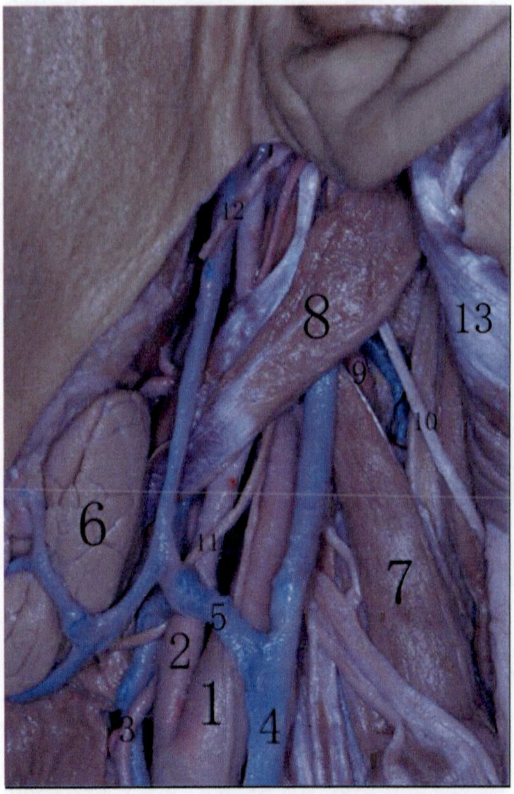

Fig. 7 Overview of left cervical region. (*1*) Internal carotid artery, (2) External carotid artery, (*3*) Superior thyroid artery, (*4*) Internal jugular vein, (*5*) Common facial common vein, (*6*) Submandibular gland, (*7*) Levator scapulae, (*8*) Digastric muscle, (*9*) Vertebral artery, (*10*) Accessory nerve, (*11*) Descending branch of hypoglossal nerve, (*12*) Marginal mandibular branch of facial nerve, (*13*) Sternocleidomastoid muscle

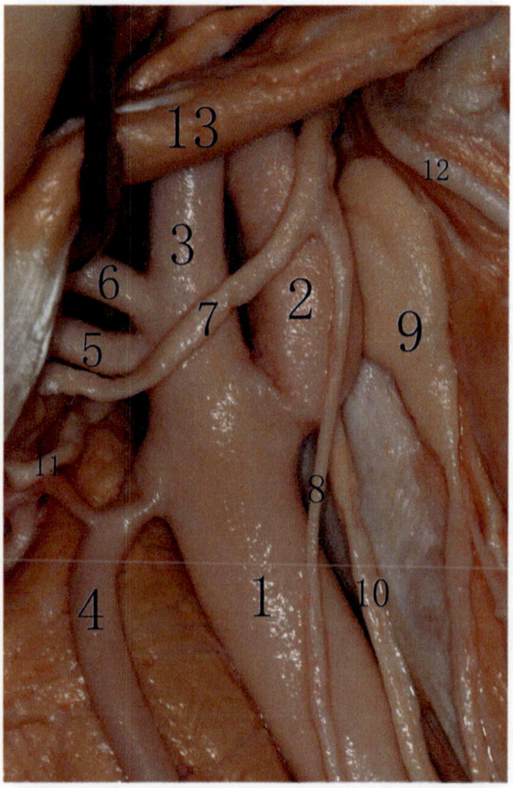

Fig. 8 Overview of left carotid complex. (*1*) Common carotid artery, (2) Internal carotid artery, (*3*) External carotid artery, (*4*) Superior thyroid artery, (*5*) Lingual artery, (*6*) Facial artery, (*7*) Hypoglossal nerve, (*8*) Descending branch of hypoglossal nerve, (9) Superior cervical sympathetic ganglion ganglia, (*10*) Vagus nerve, (*11*) Internal laryngeal nerve, (*12*) Accessory nerve, (*13*) Digastric muscle

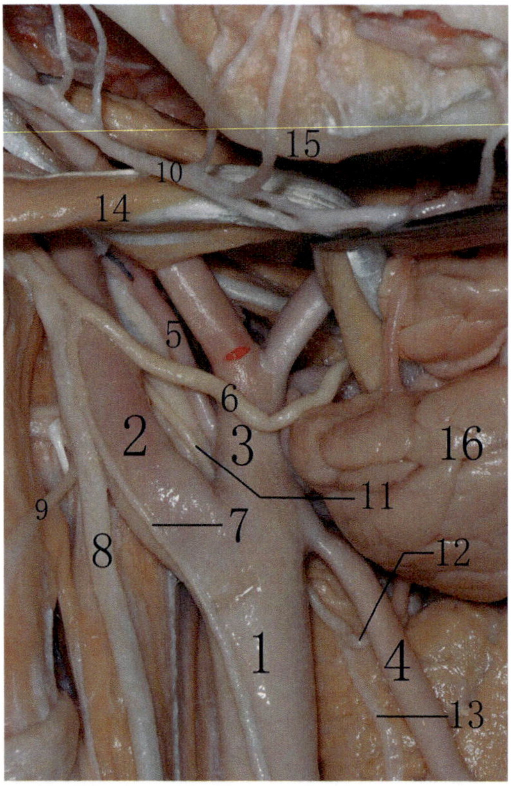

Fig. 9 Overview of right carotid complex. (*1*) Common carotid artery, (*2*) Internal carotid artery, (*3*) External carotid artery, (*4*) Superior thyroid artery, (*5*) Ascending pharyngeal artery, (*6*) Hypoglossal nerve, (*7*) Descending branch of hypoglossal nerve, (*8*) Vagus nerve, (*9*) Accessory nerve, (*10*) Marginal mandibular branch of facial nerve, (*11*) Superior laryngeal nerve, (*12*) Internal laryngeal nerve, (*13*) External laryngeal nerve, (*14*) Digastric muscle, (*15*) Mandibular angle, (*16*) Submandibular gland

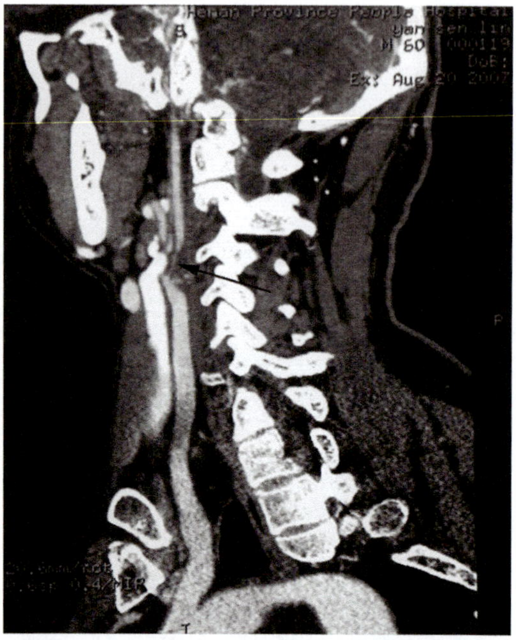

Fig. 10 Pre-op sagittal view of CT angiography revealed severe stenosis of right internal carotid artery (arrow)

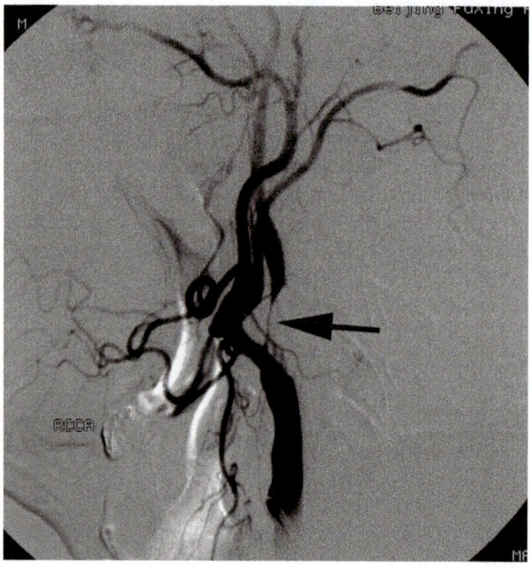

Fig. 11 Pre-op lateral view of right common carotid artery angiography study confirmed stenosis greater than 90%

Case 7.1 Carotid Endarterectomy

Clinical presentation:

A 60-year-old male presented with a 2-year history of intermittent weakness of left extremity.

Neurological exam:

Uneventful

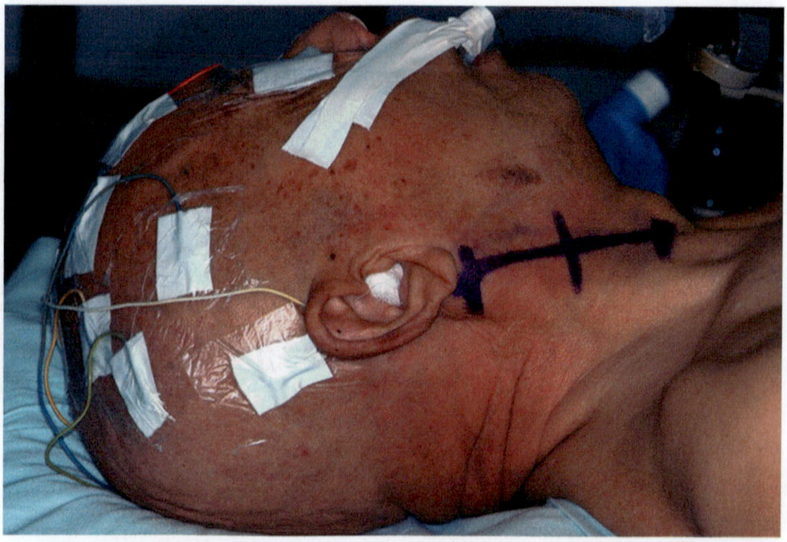

Fig. 12 The patient was placed at supine with head rotation 20° to the contralateral side. Cervical incision was marked along the medial border of sternocleidomastoid muscle from mastoid tip superiorly to cricoid cartilage inferiorly

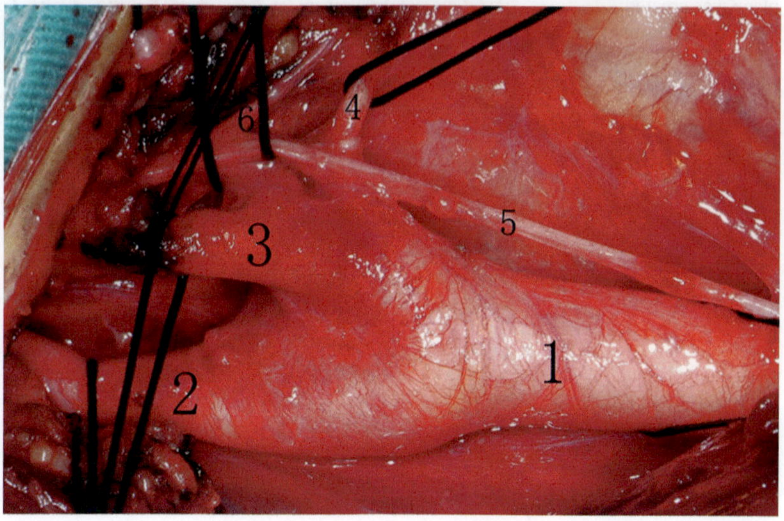

Fig. 13 Carotid sheath was opened. (*1*) Common carotid artery, (*2*) internal carotid artery, (*3*) external carotid artery, (*4*) superior thyroid artery, (*5*) descending branch of hypoglossal nerve, and (*6*) hypoglossal trunk

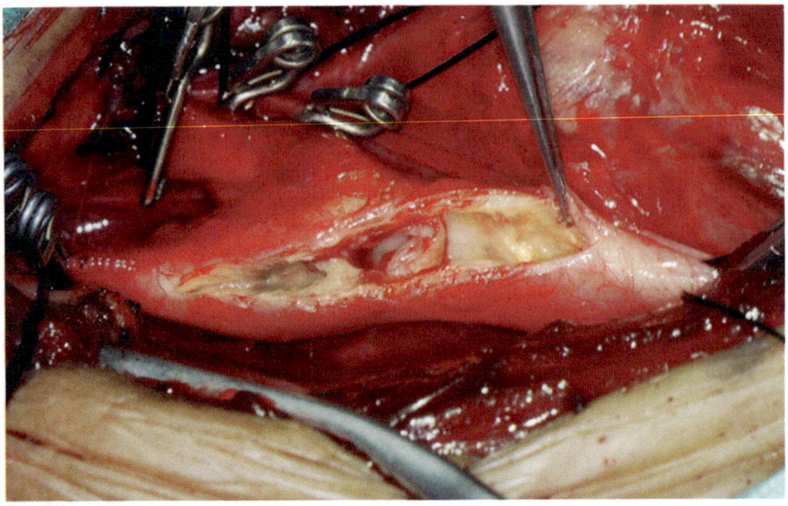

Fig. 14 The atheromatous plaque was seen after incising the anterior wall of common carotid artery and internal carotid artery

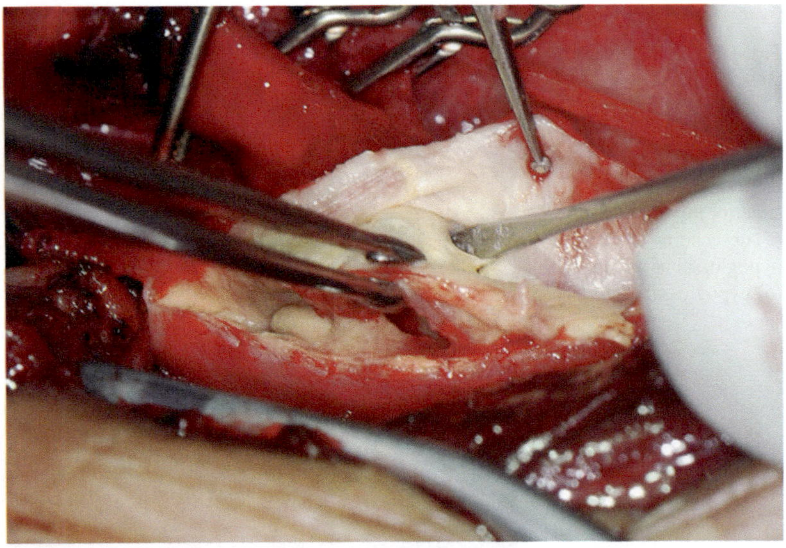

Fig. 15 The dissection plane was created between the intima and the media to remove the plaque

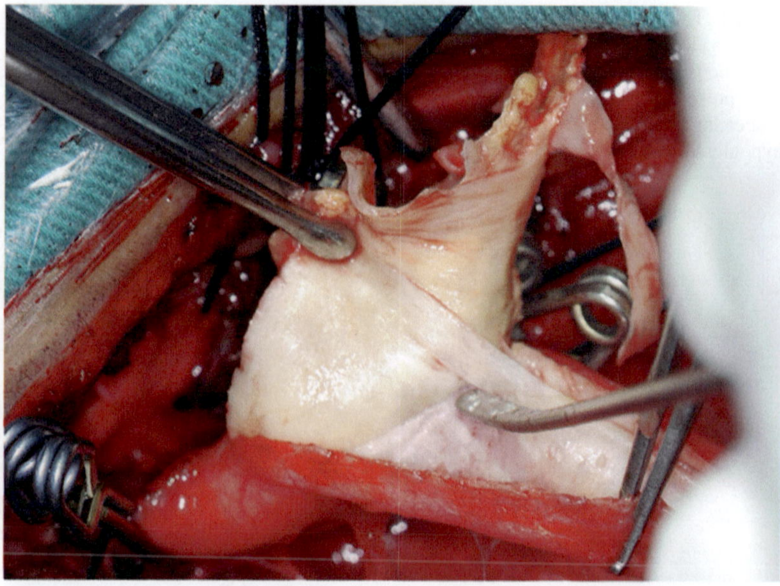

Fig. 16 The media was exposed at the back side of atheroma

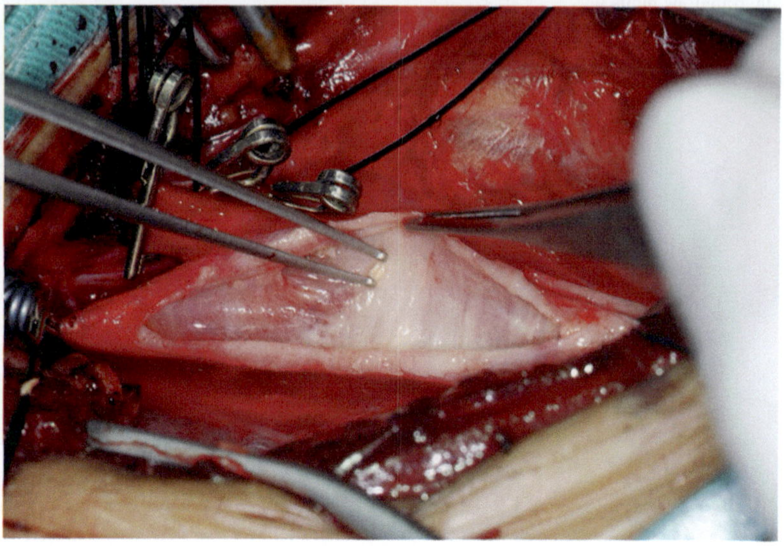

Fig. 17 The inner wall of carotid was trimmed smoothly in case of dropped clot to travel downstream

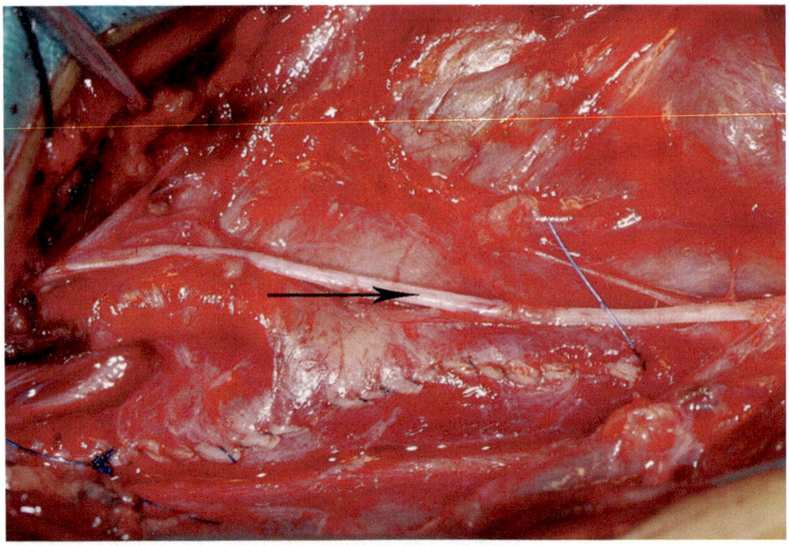

Fig. 18 The incision line was closed in a continuous fashion with good preservation of descending branch of hypoglossal nerve

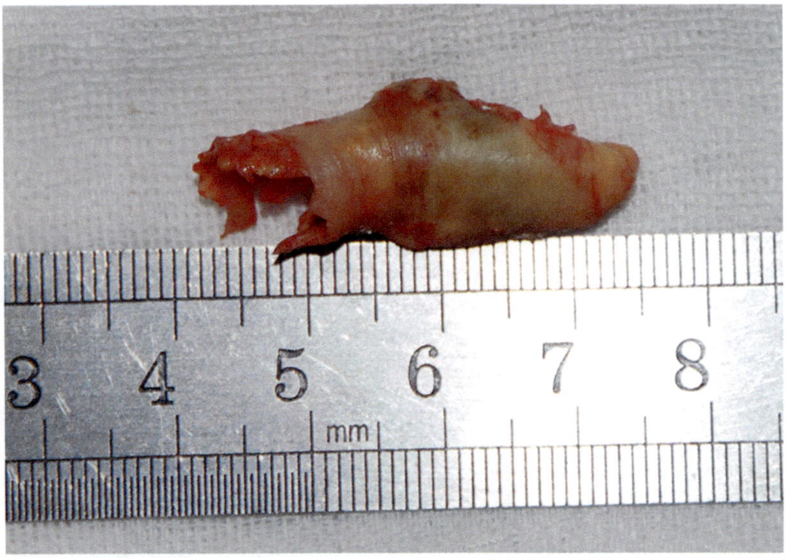

Fig. 19 The atheromatous nidus was measured longer than 3 cm

Case 7.2 Spasmodic Torticollis

Clinical presentation:

A 40-year-old lady presented with a 1 year history of right retrocollis

Neurological exam:

Right rotated and backward head position, cervical muscle stiffness.

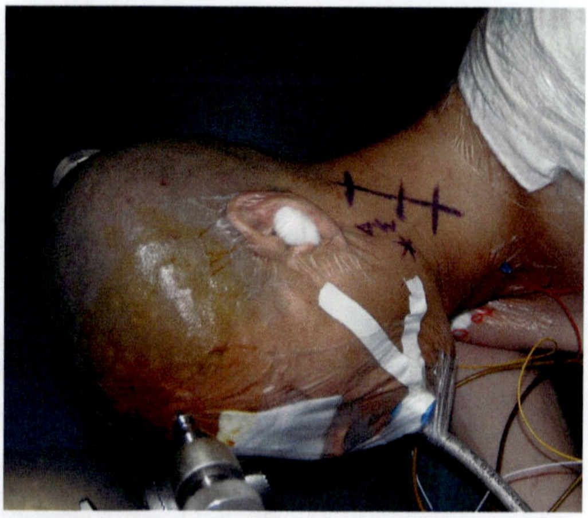

Fig. 20 The patient was placed at park-bench position. The post-sternocleidomastoid incision was labeled as exposed

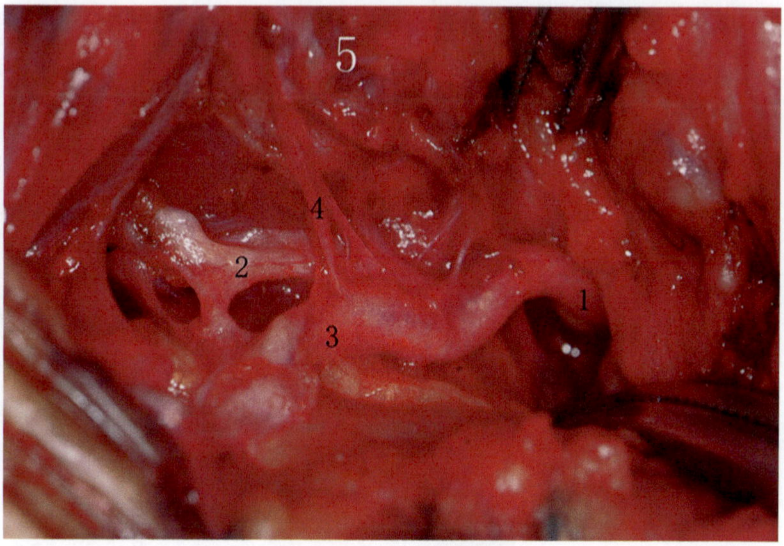

Fig. 21 The cervical fascia was severed to unveil served nerves for left sternocleidomastoid. (*1*) Accessory trunk, (*2*) sternocleidomastoid branch of accessory nerve, (*3*) trapezius branch of accessory nerve, (*4*) serving branches to sternocleidomastoid, and (*5*) sternocleidomastoid muscle

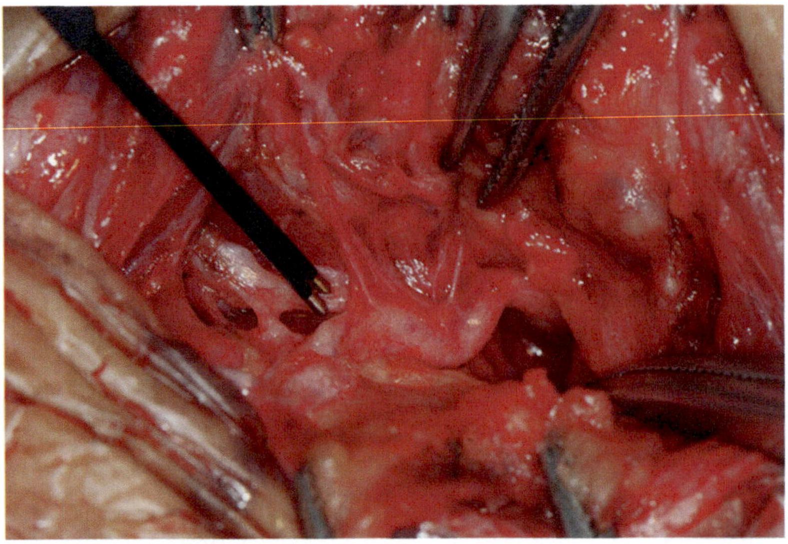

Fig. 22 The stimulator was utilized to confirm the sternocleidomastoid branch and severed following observation of muscular contraction

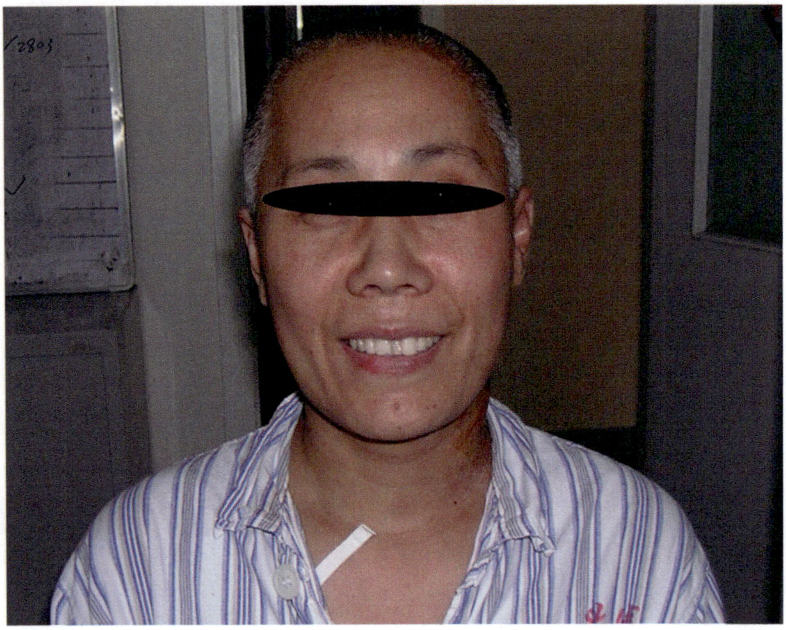

Fig. 23 Her head posture was gradually back to normal at post-op day 7

Case 7.3 Basilar Dolichoectasia

Clinical presentation:

A 55-year-old man presented with a 2-month history of left weakness and numbness.

Neurological exam:

Decreased response to gag reflex; Left muscular strength: grade IV

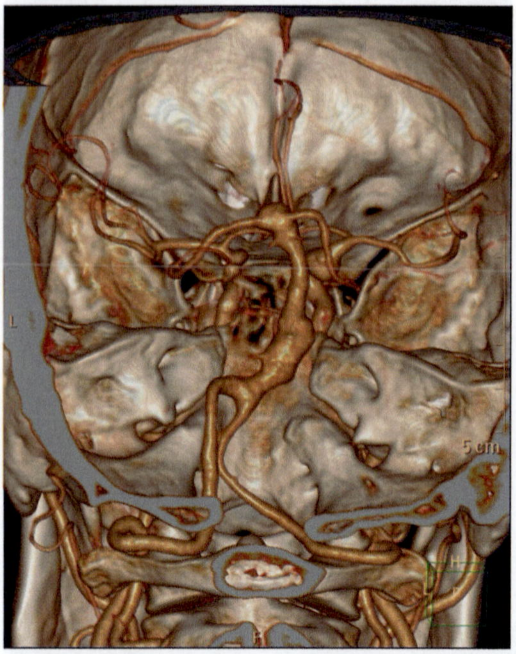

Fig. 24 Pre-op 3D CT angiography revealed a vertebral dolichoectasia and a fusiform aneurysm of right supraclinoid internal carotid artery

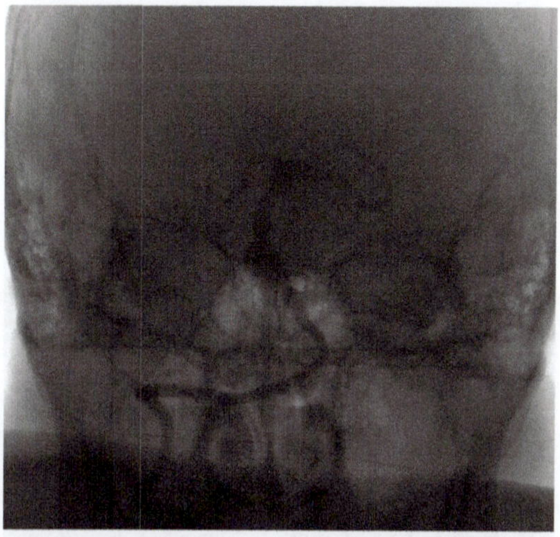

Fig. 25 Pre-op anteroposterior view of vertebrobasilar artery angiography

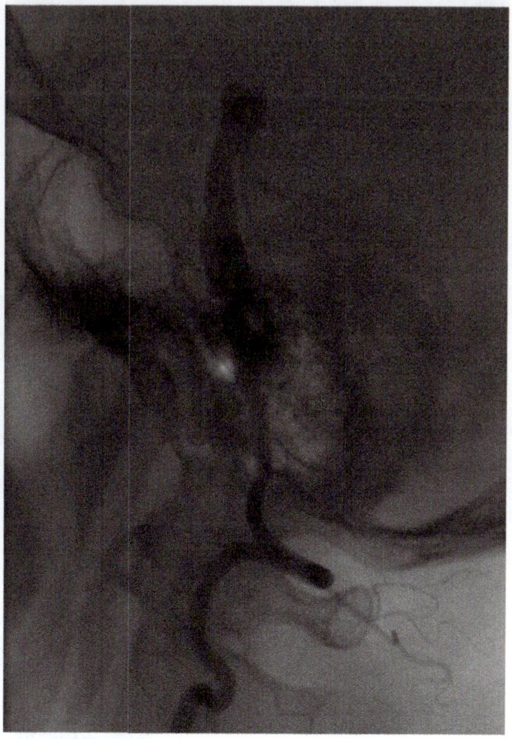

Fig. 26 Pre-op lateral view of vertebrobasilar artery angiography. Pre-op DSA showed extended dilation of vertebral trunk and multiple perforators arising from the dome

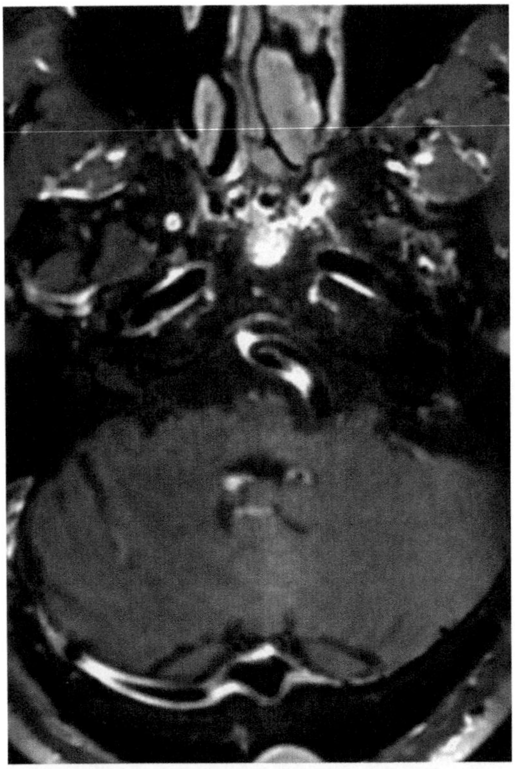

Fig. 27 Pre-op axial HR-VWI with Gd

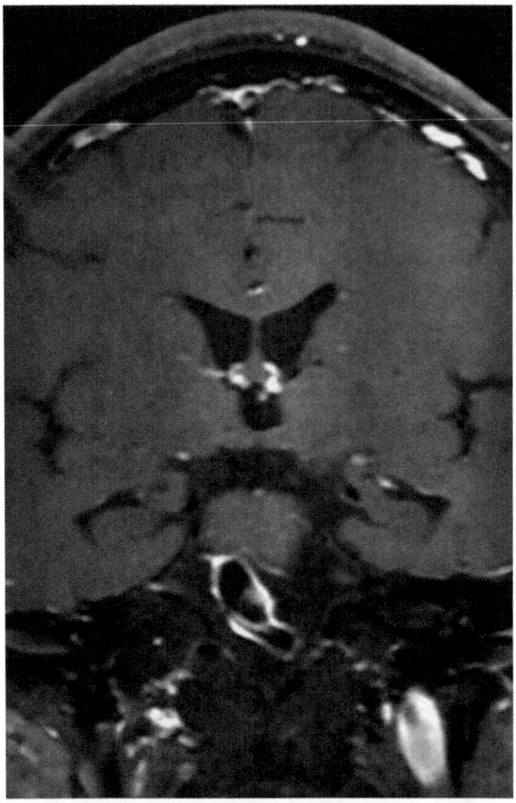

Fig. 28 Pre-op coronal HR-VWI with Gd

Fig. 29 Pre-op sagittal HR-VWI with Gd. Pre-op postcontrast HR-VWI delineated intraluminal thrombosis and thickened wall

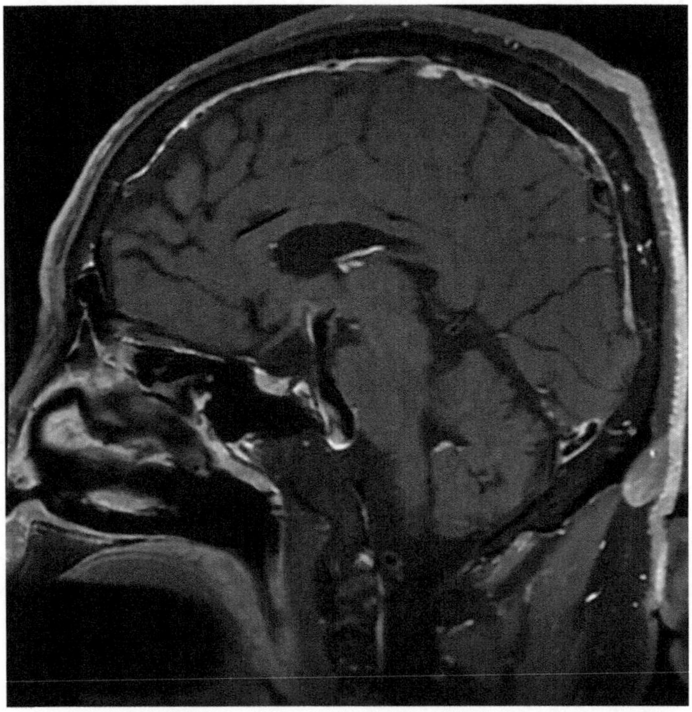

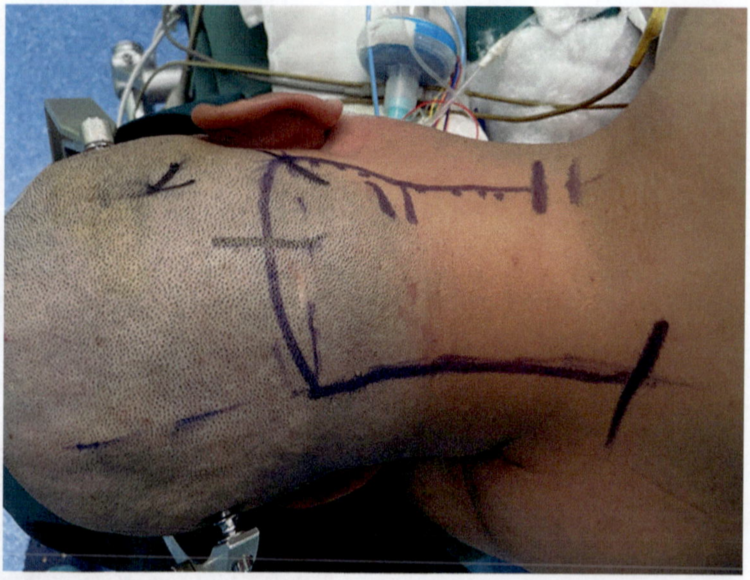

Fig. 30 The patient was placed at the left decubitus position. The standard hockey stick incision was delineated

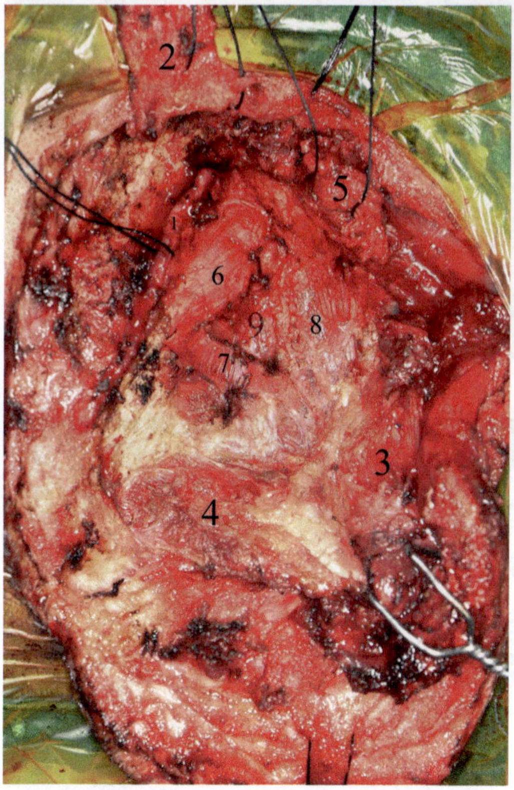

Fig. 31 The lateral suboccipital muscles were dissected in a layer-by-layer fashion to preserve all the course of occipital artery. (*1*) Intermuscular (suboccipital) segment of occipital artery, (*2*) sternocleidomastoid muscle, (*3*) splenius capitis, (*4*) semispinalis muscle, (*5*) posterior belly of digastric muscle, (*6*) Obliquus capitis superior, (*7*) rectus capitis posterior major and minor, (*8*) Obliquus capitis inferior, and (*9*) suboccipital triangle

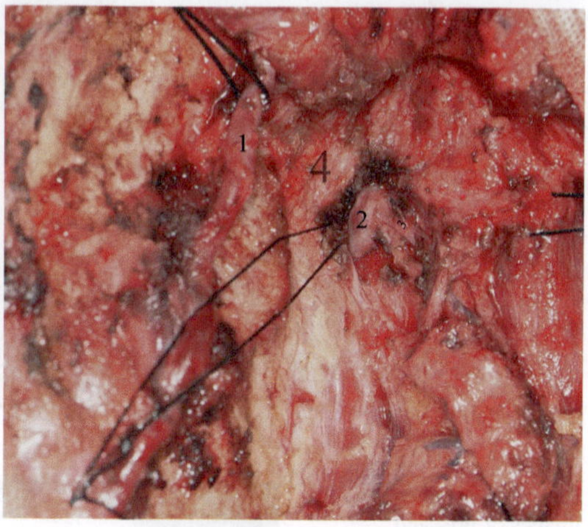

Fig. 32 The horizontal portion of V3 segment of vertebral artery was exposed after opening of suboccipital triangle. (*1*) Intermuscular (suboccipital) segment of occipital artery, (*2*) horizontal portion of V3 segment of vertebral artery, (*3*) muscular branch (artery of Salmon), and (*4*) occipital condyle

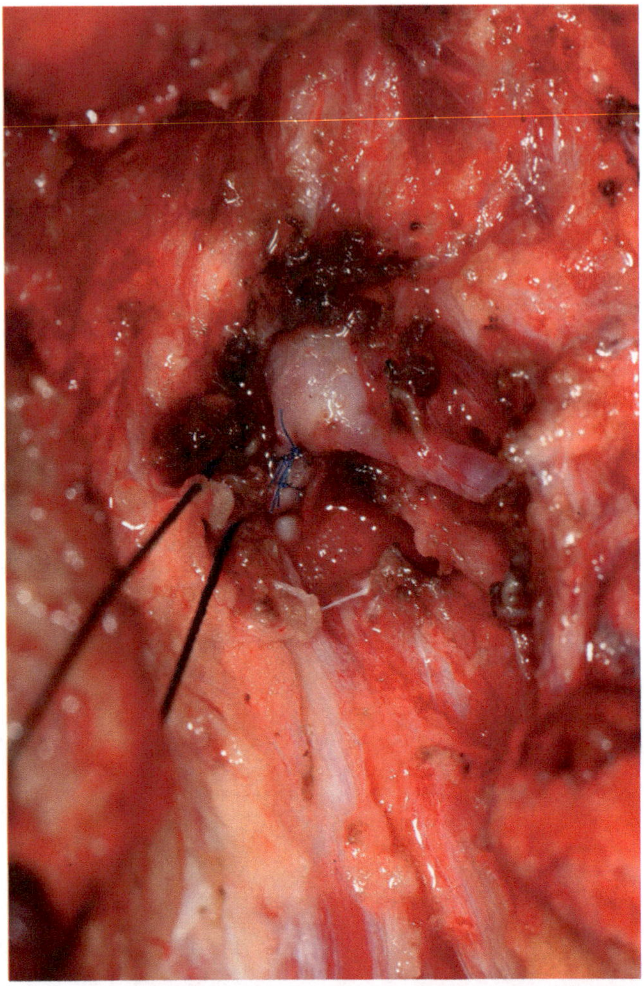

Fig. 33 The V3 was ligated by 7-0 sutures for the purpose of decompressing the inflow of dolichoectasia

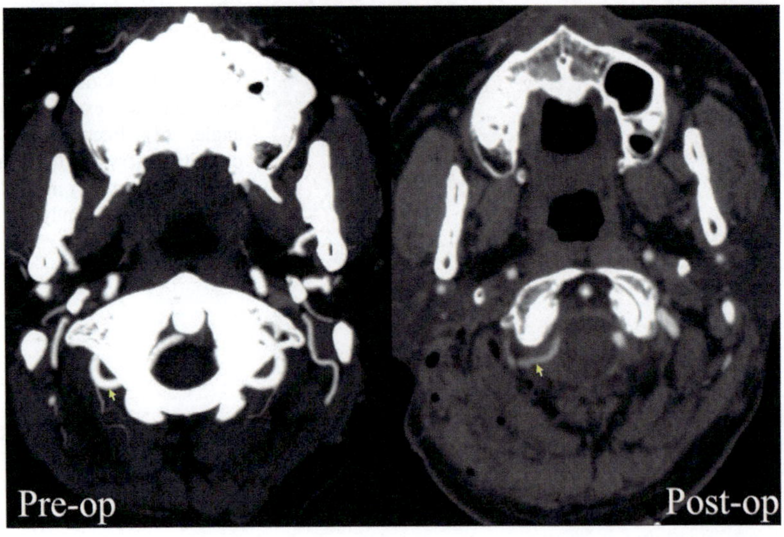

Fig. 34 Contrast images of pre- and post-op axial CT angiography

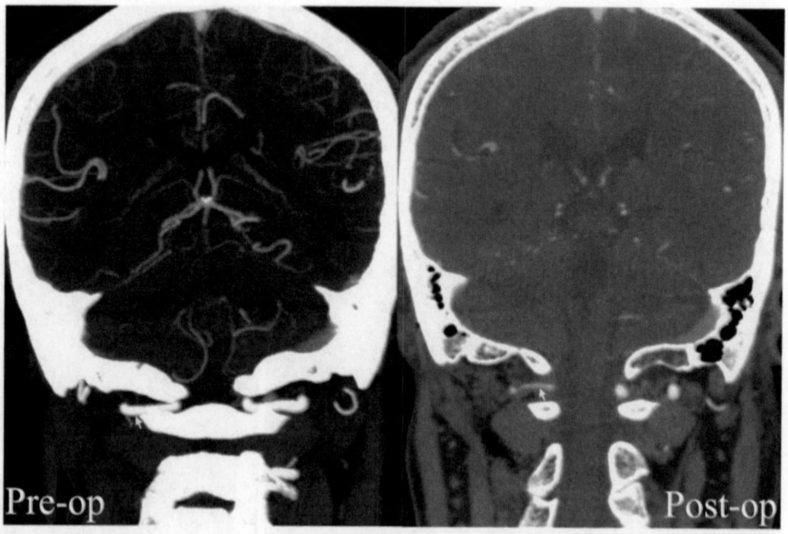

Fig. 35 Contrast images of pre- and post-op coronal CT angiography

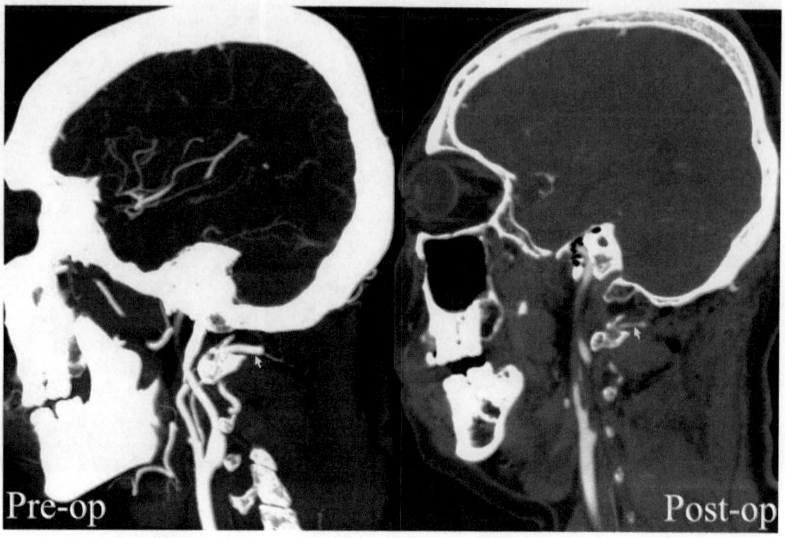

Fig. 36 Contrast images of pre- and post-op sagittal CT angiography. Pre- and post-op contrasted CT angiography revealed iatrogenic constriction of horizontal V3

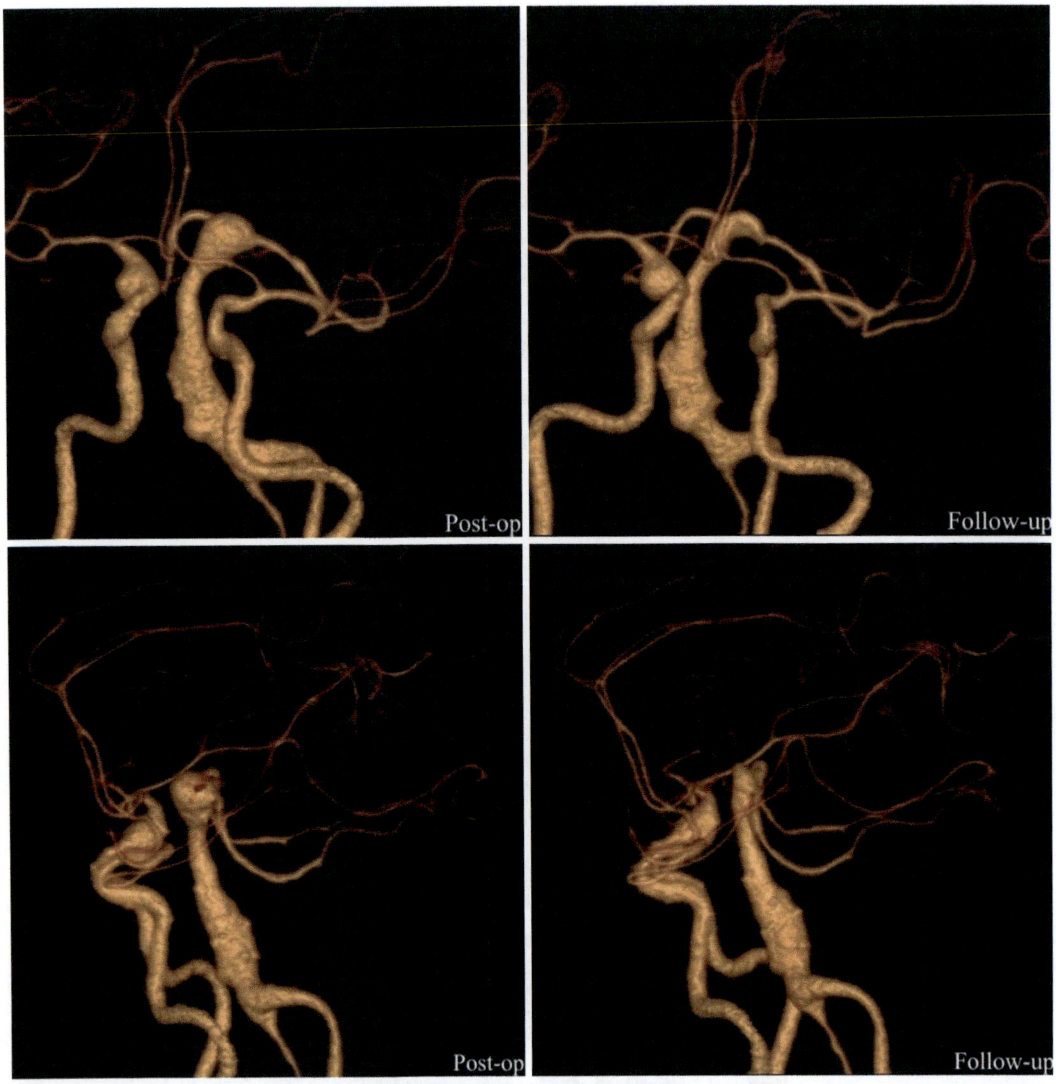

Fig. 37 3D CT angiography at 3-month follow-up revealed shrink of basilar dolichoectasia compared to the post-op status

Case 7.4 C2 Nerve Root Schwannoma

Clinical presentation:

A 42-year-old lady presented with a 1-month history of left weakness and numbness

Neurological exam:

Decreased left superficial sensation; left muscular strength: grade IV.

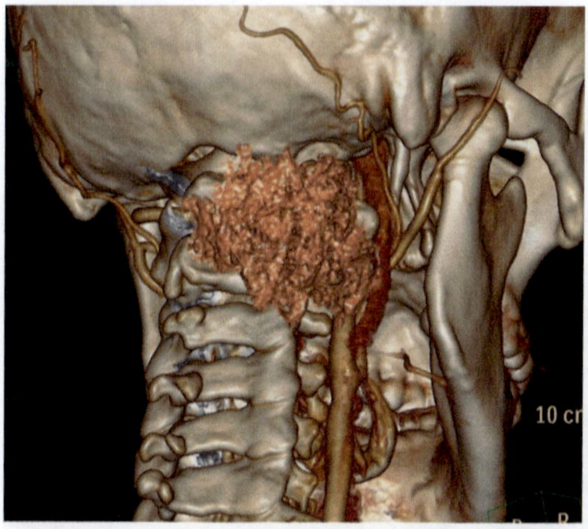

Fig. 38 Pre-op 3D CT angiography revealed mass location with surrounding structures

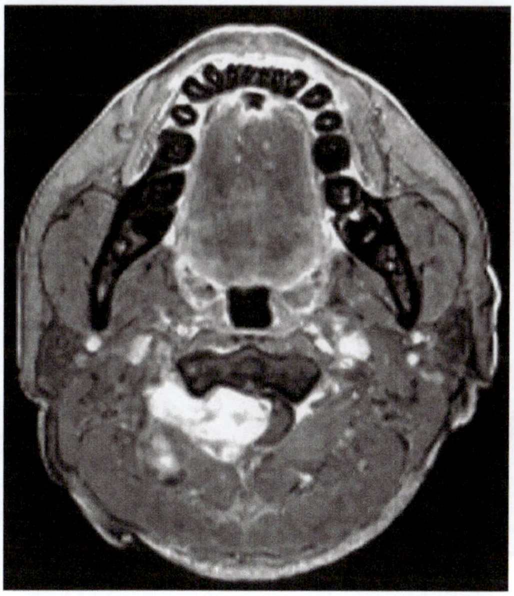

Fig. 39 Pre-op axial Gd-enhanced T1-weighted MRI

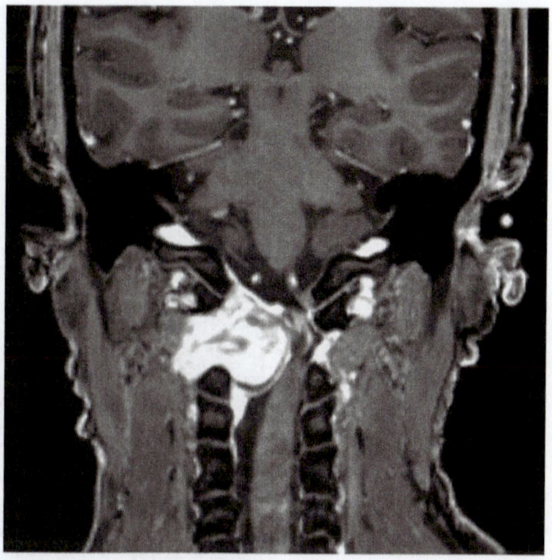

Fig. 40 Pre-op coronal Gd-enhanced T1-weighted MRI

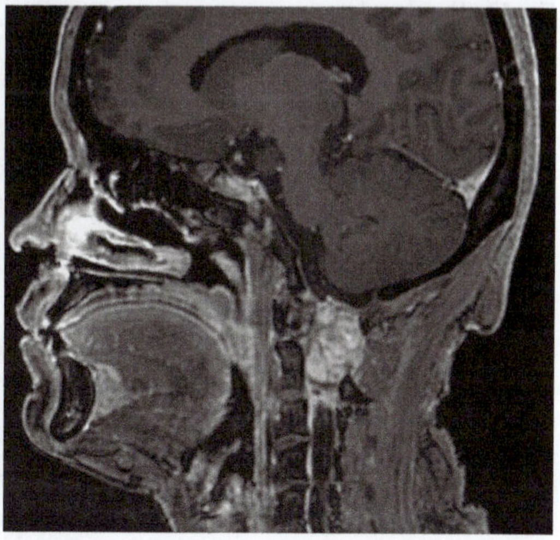

Fig. 41 Pre-op sagittal Gd-enhanced T1-weighted MRI. Pre-op Gd-enhanced T1-weighted MRI showing a dumbbell-shape lesion at right upper cervical region with bright enhancement and compressing the spinal cord laterally

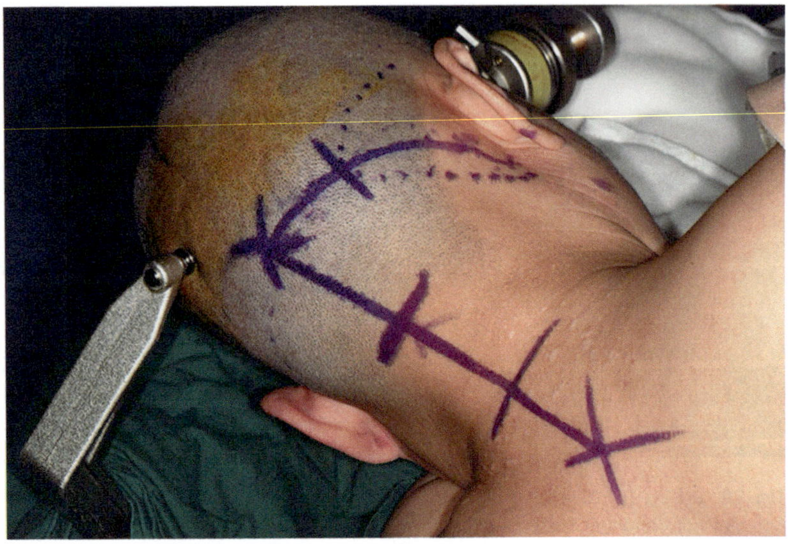

Fig. 42 The lateral decubitus position was chosen and suboccipital midline incision with lateral extension was performed

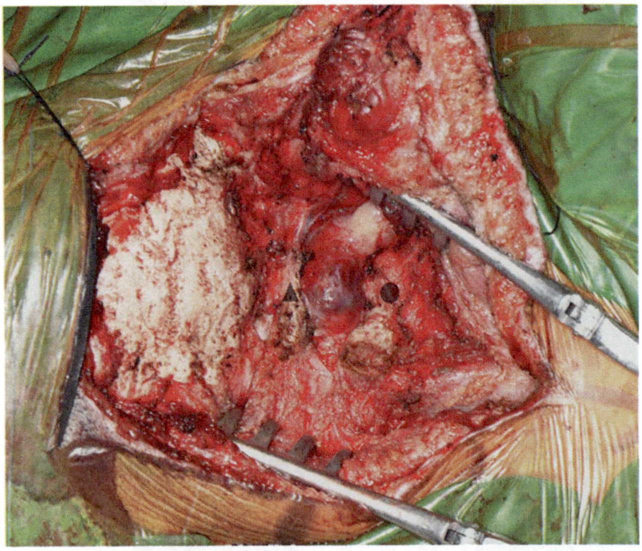

Fig. 43 The muscular flap was reflected laterally as one layer to expose squamous part of occipital bone, posterior arch of atlas, and spinous process of axis

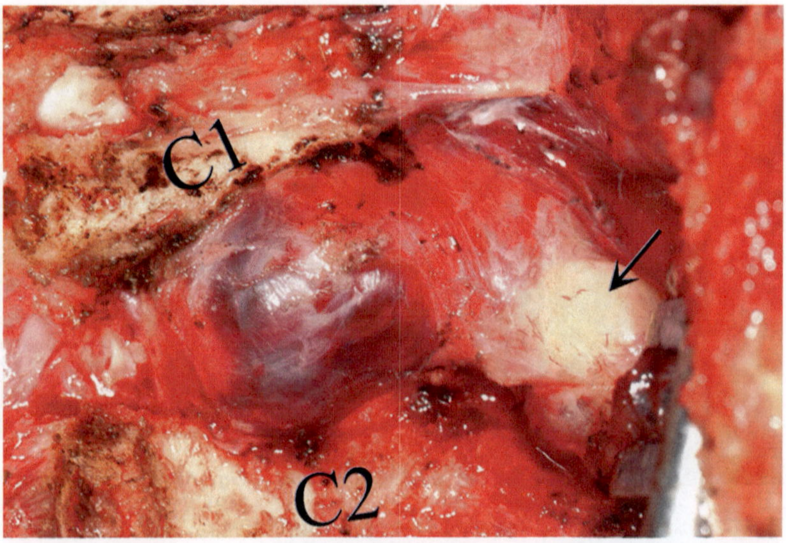

Fig. 44 The "dumbbell" hypervascularized mass was originated from the C2 root (arrow) and compressed between C1 and C2

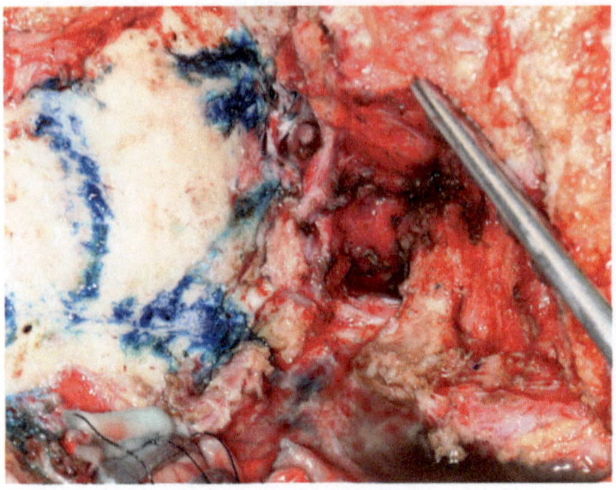

Fig. 45 The cervical dura was well preserved after tumor removal

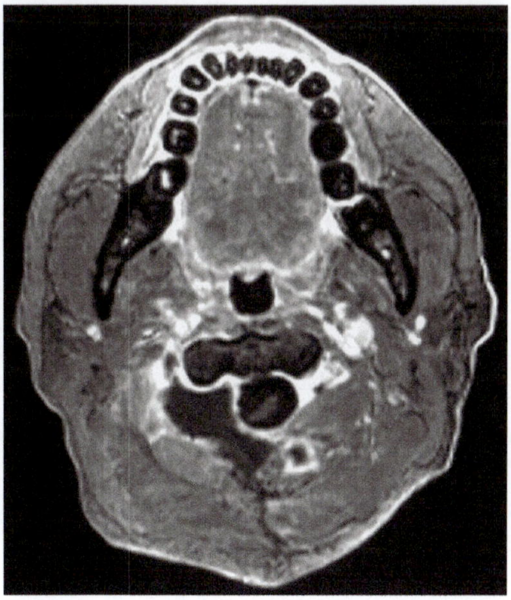

Fig. 46 Post-op axial Gd-enhanced T1-weighted MRI

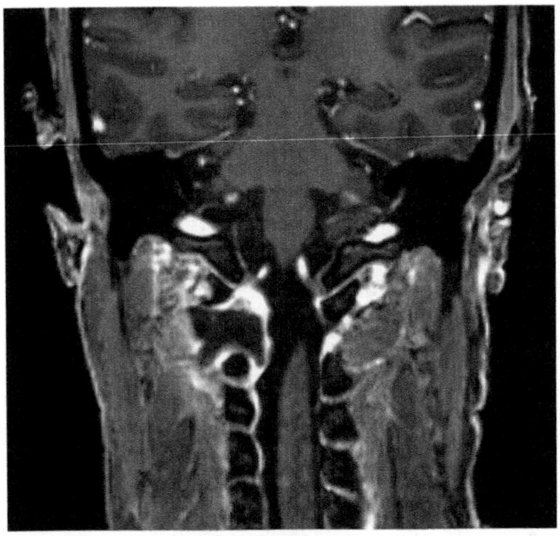

Fig. 47 Post-op coronal Gd-enhanced T1-weighted MRI

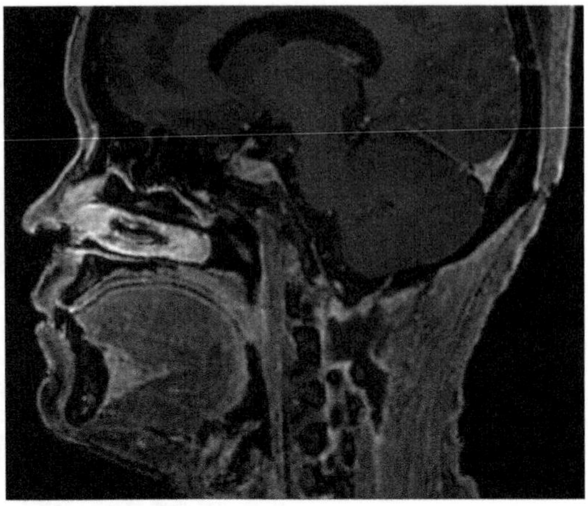

Fig. 48 Post-op sagittal Gd-enhanced T1-weighted MRI. Post-op Gd-enhanced T1-weighted MRI demonstrated total resection of C2 nerve root schwannoma without cerebrospinal leakage

Anatomy of Pineal Region and Surgical Techniques

Xiang'En Shi, Long Wang, and Hai Qian

1 Anatomy of Pineal Region

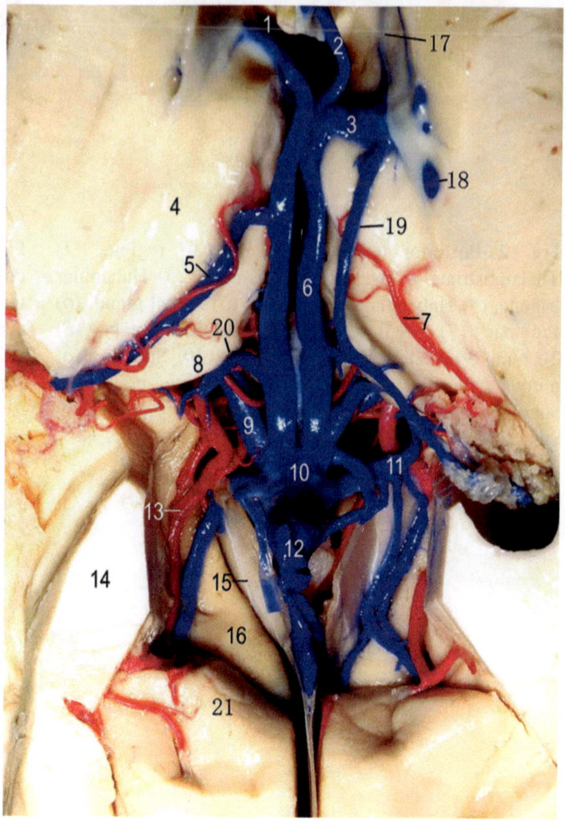

Fig. 1 Superior overview of pineal region. (*1*) Foramen of Monro, (*2*) Septal vein, (*3*) Thalamostriate vein, (*4*) Thalamus, (*5*) Superior choroidal vein, (*6*) Internal cerebral vein, (*7*) Lateral posterior choroidal artery, (*8*) Pulvinar, (*9*) Basal vein of Rosenthal, (*10*) Vein of Galen, (*11*) Medial occipital vein, (*12*) Straight sinus, (*13*) Posterior cerebral artery, (*14*) Splenium of corpus callosum, (*15*) Tentorium, (*16*) Isthmus of cingulate gyrus, (*17*) Anterior caudate vein, (*18*) Posterior caudate vein, (*19*) Superior choroidal vein, (*20*) Medial atrial vein, (*21*) Cuneus

X. Shi (✉) · L. Wang · H. Qian
Neurological Surgery, SanBo Brain Hospital, Capital Medical University, Beijing, China
e-mail: shixen@ccmu.edu.cn; wanglong@ccmu.edu.cn; qianh@ccmu.edu.cn

X. Shi et al. (eds.), *Microneurosurgical Anatomy and Surgical Technique*, https://doi.org/10.1007/978-981-19-8273-6_8

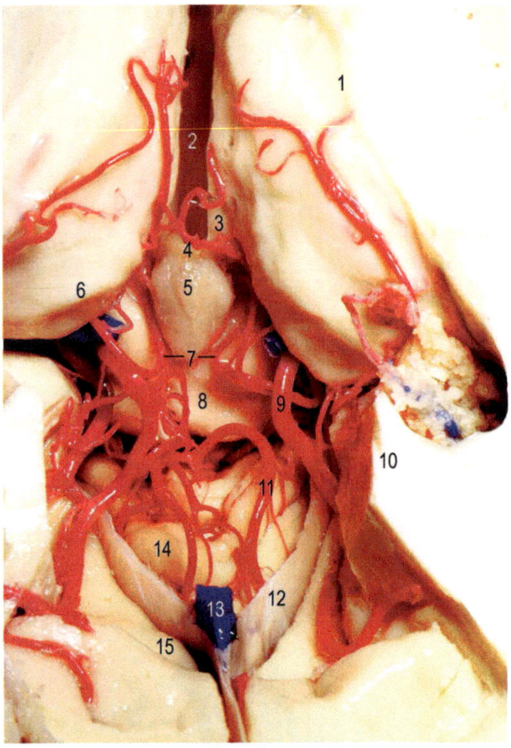

Fig. 2 Posterior overview of pineal region. (*1*) Thalamostriate groove, (*2*) 3rd ventricle, (*3*) Habenular trigone, (*4*) Habenular commissure, (*5*) Pineal gland, (*6*) Pulvinar, (*7*) Medial posterior choroidal artery, (*8*) Tectum, (*9*) Posterior cerebral artery, (*10*) Splenium of corpus callosum, (*11*) Medial hemispheric branch of superior cerebellar artery, (*12*) Tentorium, (*13*) Straight sinus, (*14*) Culmen, (*15*) Isthmus of cingulate gyrus

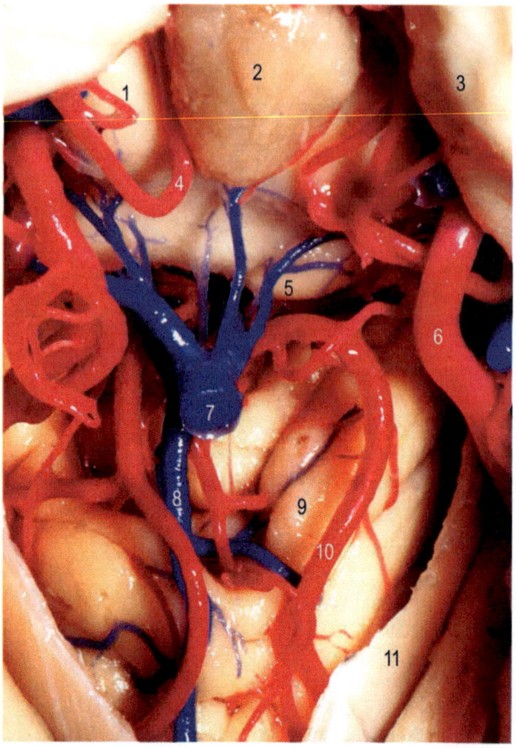

Fig. 3 Overview of quadrigeminal cistern. (*1*) Superior colliculus, (*2*) Pineal gland, (*3*) Pulvinar, (*4*) Medial posterior choroidal artery, (*5*) Inferior colliculus, (*6*) Posterior cerebral artery, (*7*) Vein of Galen, (*8*) Superior vermian vein, (*9*) Central lobe, (*10*) Medial hemispheric branch of superior cerebellar artery, (*11*) Tentorium

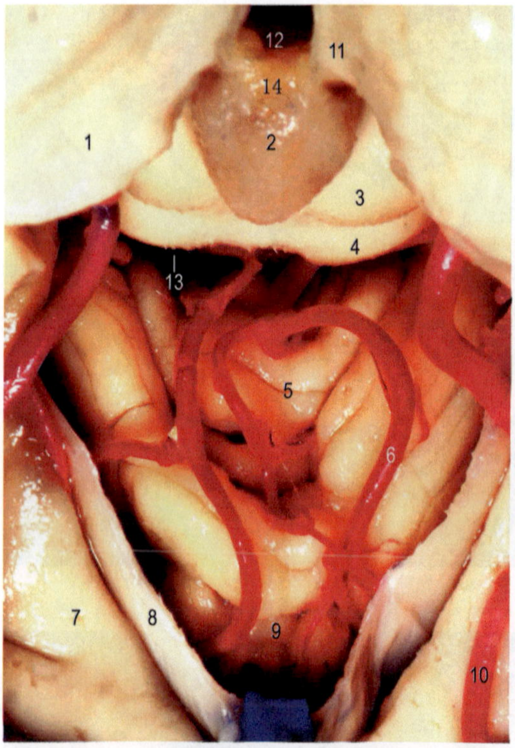

Fig. 4 Posterior view of posterior wall of 3rd ventricle. (*1*) Pulvinar, (*2*) Pineal gland, (*3*) Superior colliculus, (*4*) Inferior colliculus, (*5*) Central lobe, (*6*) Medial hemispheric branch of superior cerebellar artery, (*7*) Isthmus of cingulated gyrus, (*8*) Tentorium, (*9*) Declive, (*10*) Posterior cerebral artery, (*11*) Habenular trigone, (*12*) Pineal recess, (*13*) Cerebellomesencephalic fissure, (*14*) Habenular commissure

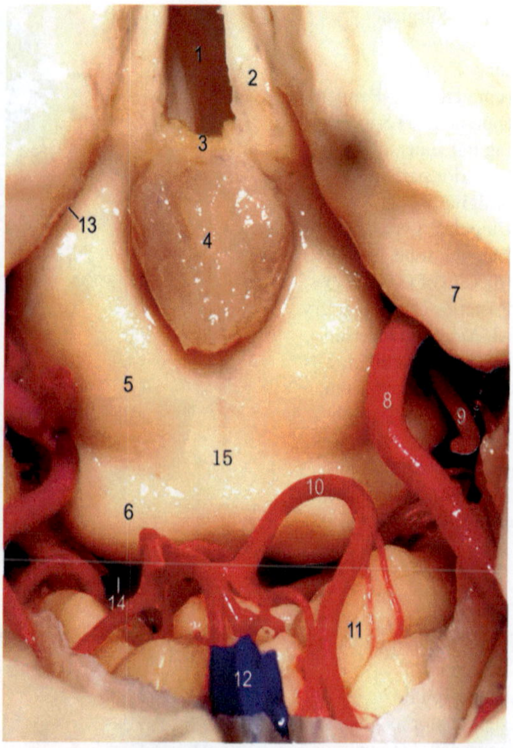

Fig. 5 Posterior view of pineal region. (*1*) 3rd ventricle, (*2*) Habenular trigone, (*3*) Habenular commissure, (*4*) Pineal gland, (*5*) Superior colliculus, (*6*) Inferior colliculus, (*7*) Right cistern part of pulvinar, (*8*) Posterior cerebral artery, (*9*) Medial posterior choroidal artery, (*10*) Medial hemispheric branch of superior cerebellar artery, (*11*) Culmen, (*12*) Straight sinus, (*13*) Left cistern part of pulvinar, (*14*) Cerebellomesencephalic fissure, (*15*) Cruciform sulcus

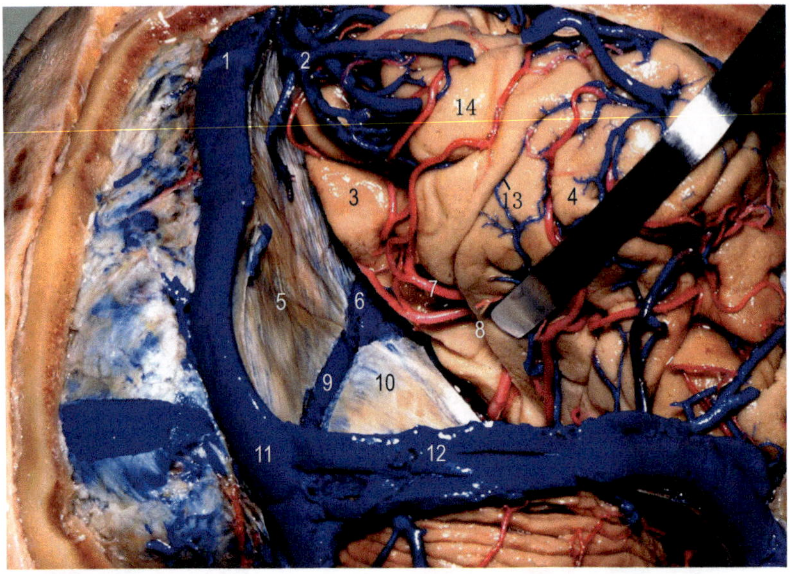

Fig. 6 Surgical trajectory of suboccipital-supratentorial (Poppen) approach. (*1*) Superior sagittal sinus, (*2*) Posterior parietal bridging vein, (*3*) Cingulate gyrus, (*4*) Lingular lobe, (*5*) Falx, (*6*) Vein of Galen, (*7*) Posterior cerebral artery, (*8*) Anterior calcarine sulcus, (*9*) Straight sinus, (*10*) Tentorium, (*11*) Torcular herophili (Confluence of sinuses), (*12*) Transverse sinus, (*13*) Posterior calcarine sulus, (*14*) Cuneus

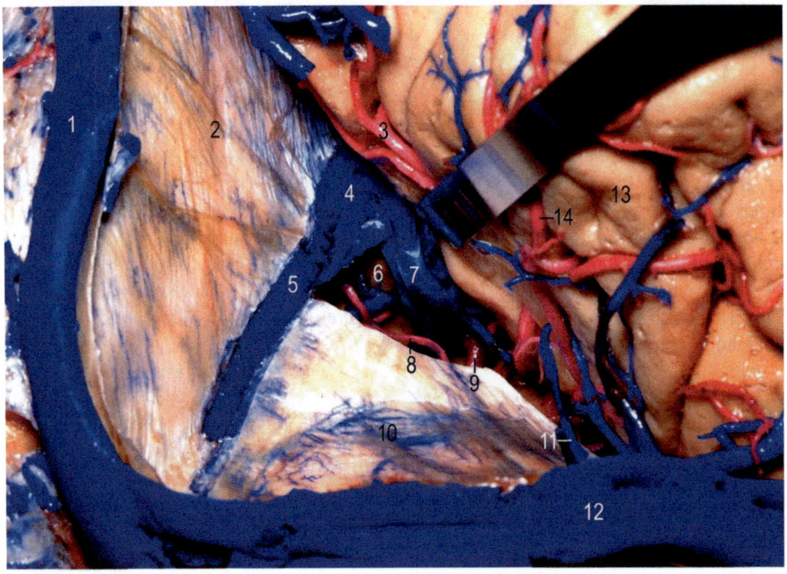

Fig. 7 Enlarged view of trajectory of suboccipital-supratentorial (Poppen) approach. (*1*) Superior sagittal sinus, (*2*) Falx, (*3*) Posterior cerebral artery, (*4*) Vein of Galen, (*5*) Straight sinus, (*6*) Tectum, (*7*) Internal cerebral artery, (*8*) Superior cerebellar artery, (*9*) Posterior cerebral artery, (*10*) Tentorium, (*11*) Occipital bridging vein, (*12*) Transverse sinus, (*13*) Lingular lobe, (*14*) Posterior inferior temporal artery

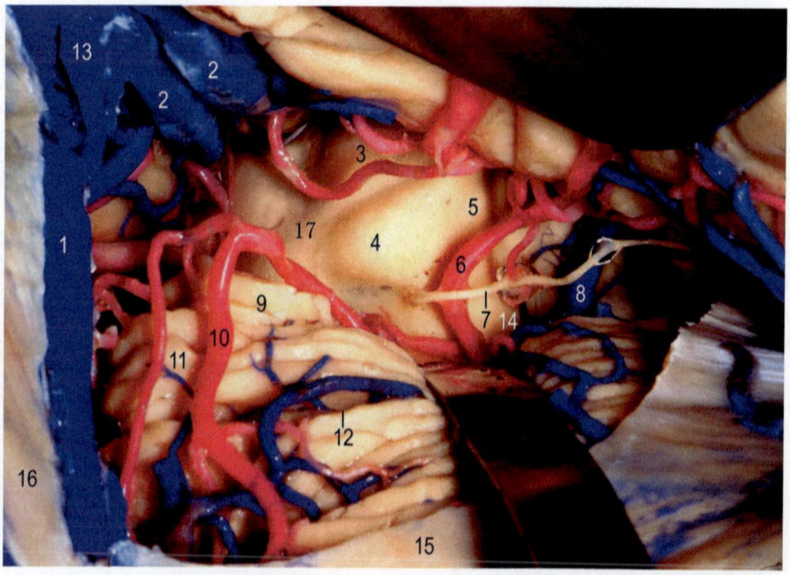

Fig. 8 The trajectory of suboccipital-transtentorial approach. (*1*) Straight sinus, (*2*) Left: Internal cerebral vein; Right: Basal vein of Rosenthal, (*3*) Superior colliculus, (*4*) Inferior colliculus, (*5*) Medial geniculate body, (*6*) Cerebellomesencephalic segment of superior cerebellar artery, (*7*) Trochlear nerve, (*8*) Lateral mesencephalic vein, (*9*) Culmen, (*10*) Medial hemispheric branch of superior cerebellar artery, (*11*) Declive, (*12*) Primary fissure, (*13*) Superior cerebellar vein, (*14*) Superior cerebellar peduncle, (*15*) Tentorium, (*16*) Falx, (*17*) Cruciform sulcus

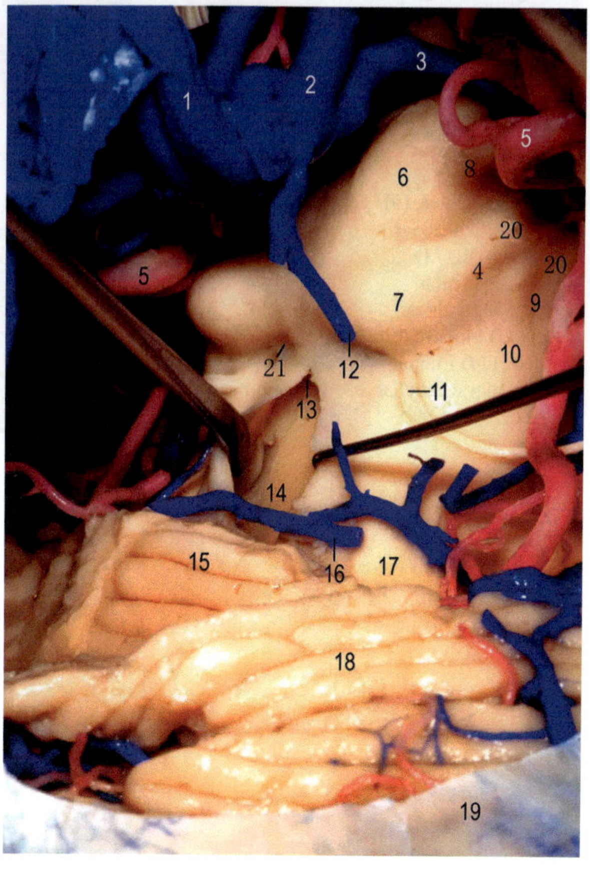

Fig. 9 Posterolateral view of dorsal mesencephalon. (*1*) Vein of Galen, (*2*) Internal cerebral vein, (*3*) Basal vein of Rosenthal, (*4*) Brachium colliculi inferioris, (*5*) Posterior cerebral artery, (*6*) Superior colliculus, (*7*) Inferior colliculus, (*8*) Brachium colliculi superioris, (*9*) Lemniscal trigone, (*10*) Superior cerebellar peduncle, (*11*) Trochlear nerve, (*12*) Tectal vein, (*13*) Cerebral aqueduct, (*14*) 4th ventricle, (*15*) Declive, (*16*) Vein of superior cerebellar peduncle, (*17*) Middle cerebellar peduncle, (*18*) Quadrangular lobule, (*19*) Tentorium, (*20*) Medial geniculate body, (*21*) Frenulum veli

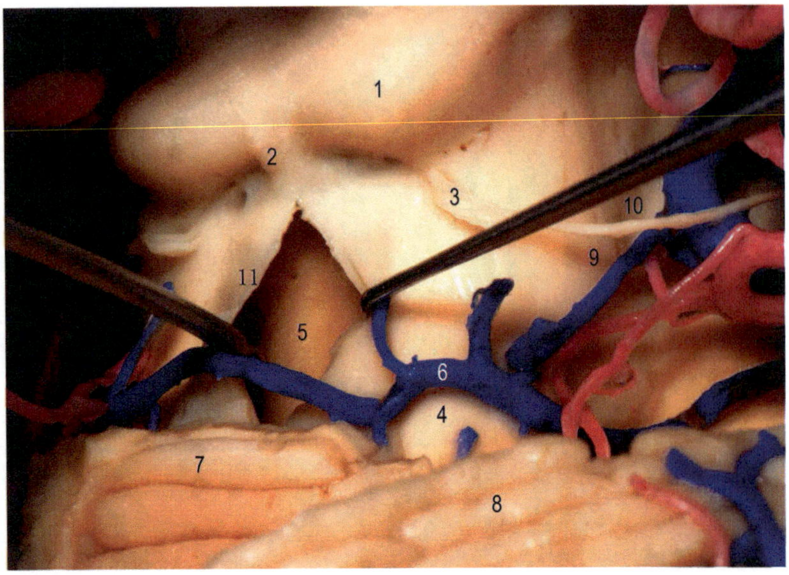

Fig. 10 Opening the upper roof of 4th ventricle. (*1*) Inferior colliculus, (*2*) Frenulum veli, (*3*) Trochlear nerve, (*4*) Superior cerebellar peduncle, (*5*) Floor of 4th ventricle, (*6*) Vein of superior cerebellar peduncle, (*7*) Central lobe, (*8*) Quadrangular lobule, (*9*) Superior cerebellar peduncle, (*10*) Lemniscal trigone, (*11*) Superior medullary velum

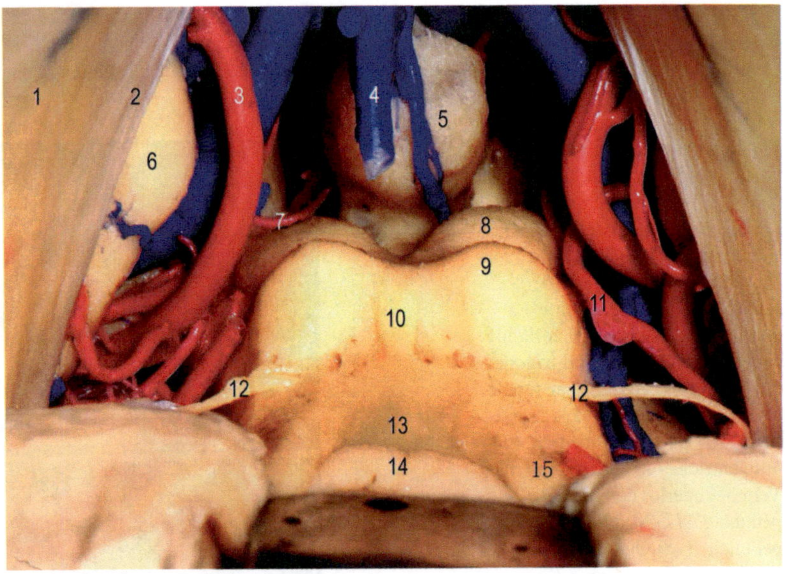

Fig. 11 Posteroinferior view of pineal region. (*1*) Tentorium, (*2*) Tentorial edge, (*3*) Quadrigeminal segment of posterior cerebral artery, (*4*) Superior cerebellar vein, (*5*) Pineal gland, (*6*) Parahippocampal gyrus, (*7*) Medial posterior choroidal artery, (*8*) Superior colliculus, (*9*) Inferior colliculus, (*10*) Frenulum veli, (*11*) Superior cerebellar artery, (*12*) Trochlear nerve, (*13*) Floor of 4th ventricle, (*14*) Lingula, (*15*) Superior cerebellar peduncle

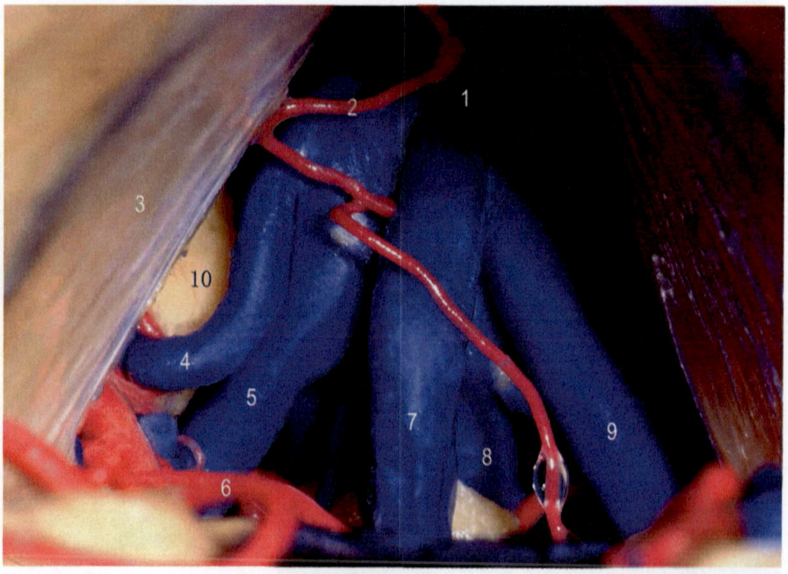

Fig. 12 Posterior view of deep cerebral veins of pineal region. (*1*) Vein of Galen, (*2*) Medial tentorial artery, (*3*) Tentorium, (*4*) Internal occipital vein, (*5*) Left basal vein of Rosenthal, (*6*) Lateral hemispheric branch of superior cerebellar artery, (*7*) Superior cerebellar vein, (*8*) Internal cerebral vein, (*9*) Right basal vein of Rosenthal, (*10*) Cistern part of pulvinar

Case 8.1 Vein of Galen Malformation

Clinical presentation:

An 18-month boy presented with a 6-month history of gait instability.

Neurological exam:

Truncal ataxia

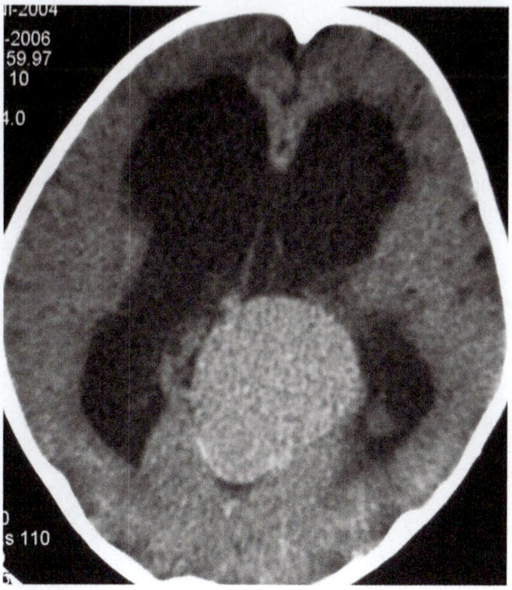

Fig. 13 Pre-op axial CT revealed a giant, well-marginated globular mass involving posterior wall of 3rd ventricle with hyperdense signal

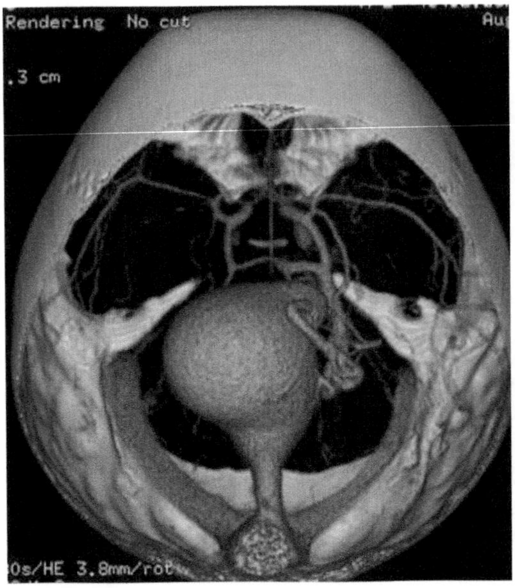

Fig. 14 Superior view of 3D CT angiography

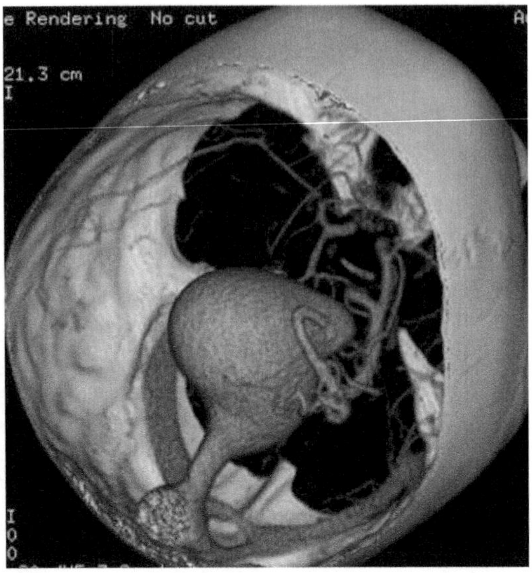

Fig. 15 Superolateral view of 3D CT angiography. Pre-op 3D CT angiography delineated the venous dilatation of the vein of Galen. Posterior cerebral artery was identified as nodual feeders

Fig. 16 Pre-op axial T2-weighted MRI illuminated black flow void indicating nidus

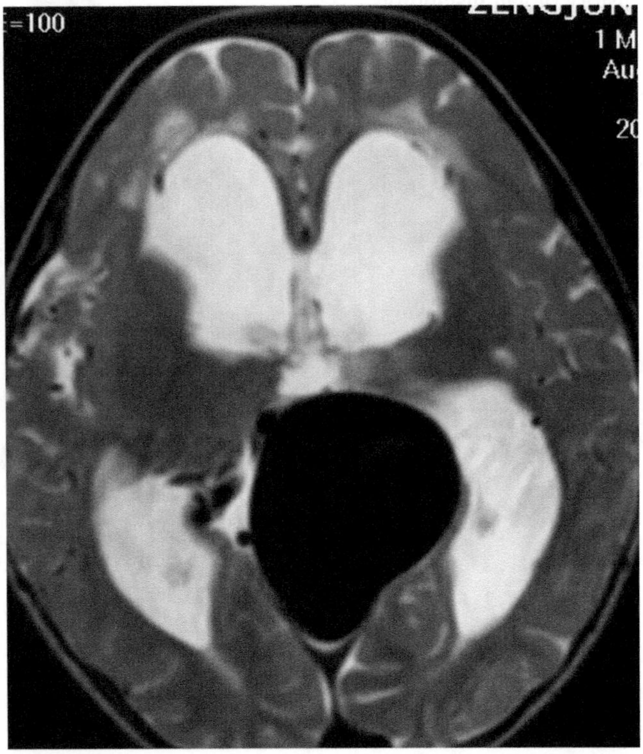

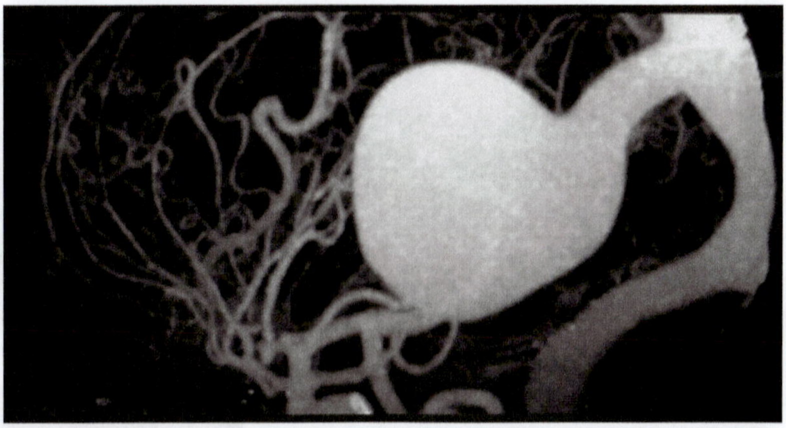

Fig. 17 Pre-op lateral view of MR venography

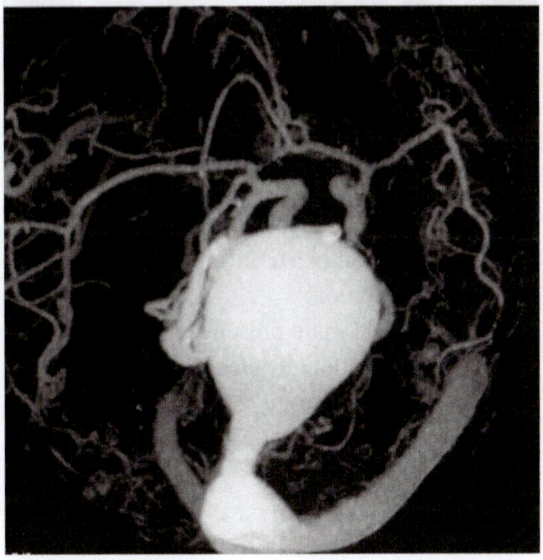

Fig. 18 Pre-op axial view of MR venography. Pre-op MR venography indicated aberrant dilation of the vein of Galen that outflow into straight sinus

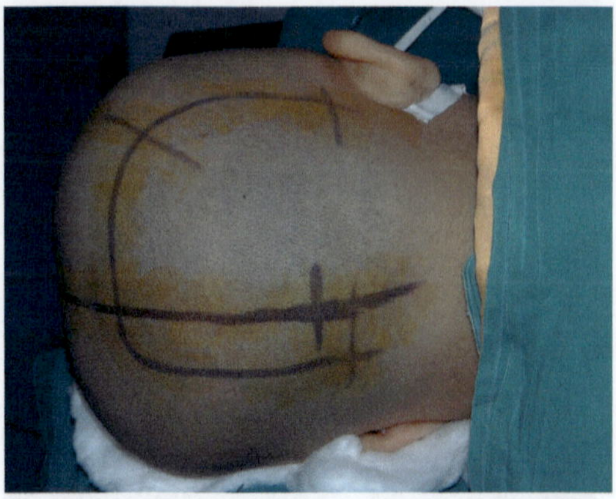

Fig. 19 The child was placed in lateral decubitus. A "horseshoe" shape incision was made for subsequent right parietooccipital craniotomy with 1cm off the midline

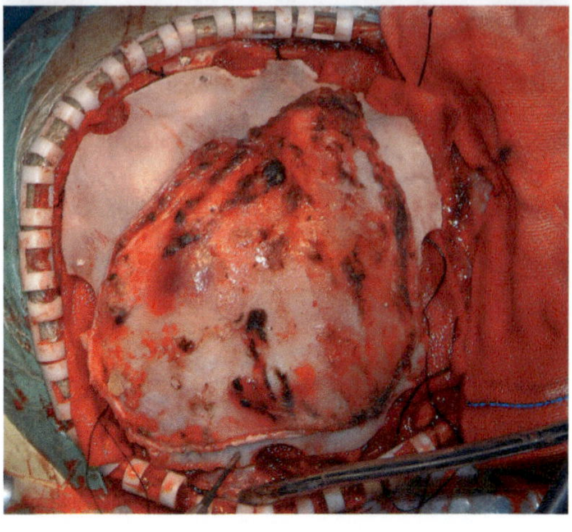

Fig. 20 The pericranium was dissected and preserved

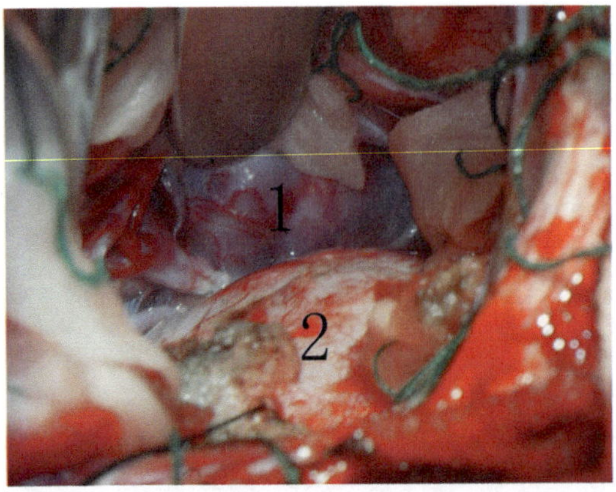

Fig. 21 The occipital parenchyma was elevated and venous malformation was visualized and clip ligated. (*1*) Vein of Galen malformation and (*2*) tentorium

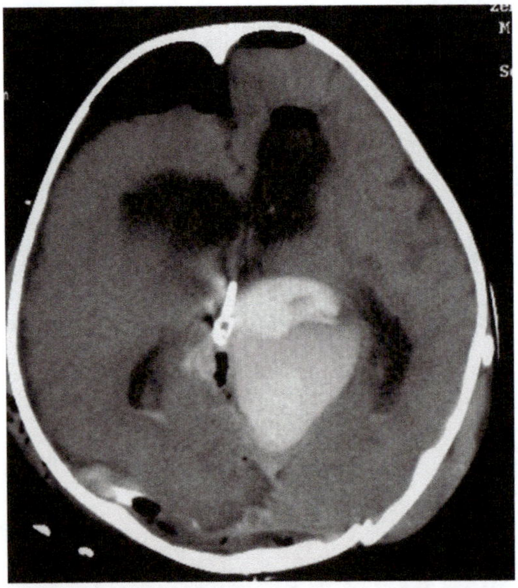

Fig. 22 Post-op CT revealed intraluminal thrombosis

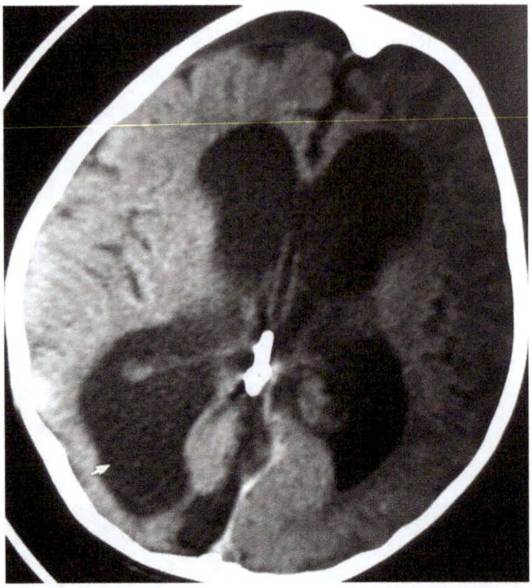

Fig. 23 Axial CT scan at 1-year follow-up showing the disappearance of giant malformation

Case 8.2 Cerebellar Medulloblastoma

Clinical presentation:

A 17-year-old male presented with a 2-month history of intermittent headache.

Neurological exam:

Papilledema

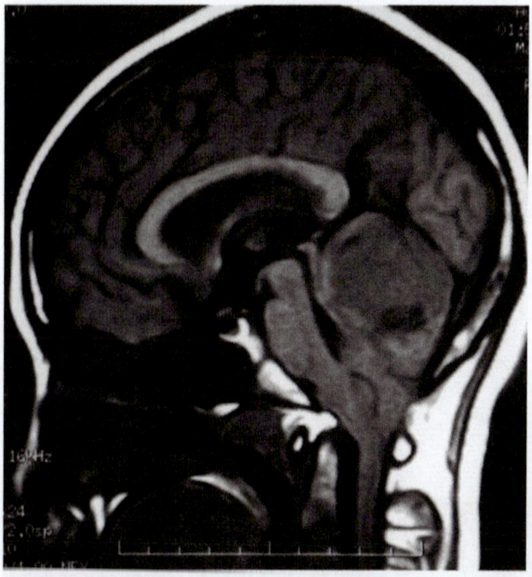

Fig. 24 Pre-op sagittal T1-weighted MRI

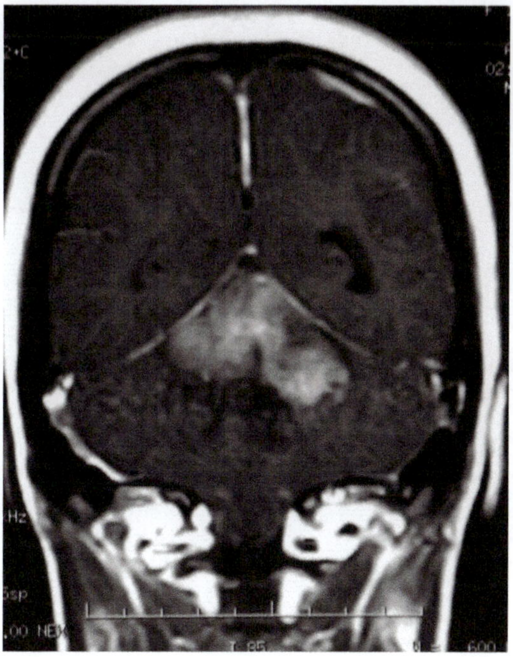

Fig. 25 Pre-op coronal Gd-enhanced T1-weighted MRI

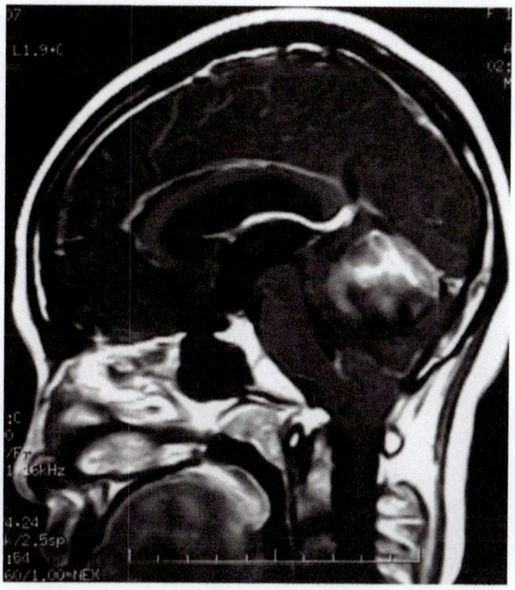

Fig. 26 Pre-op sagittal Gd-enhanced T1-weighted MRI. Pre-op MRI showed a well-marginated mass of cerebellar vermis with cystic component. The lesion was heterogeneously enhanced

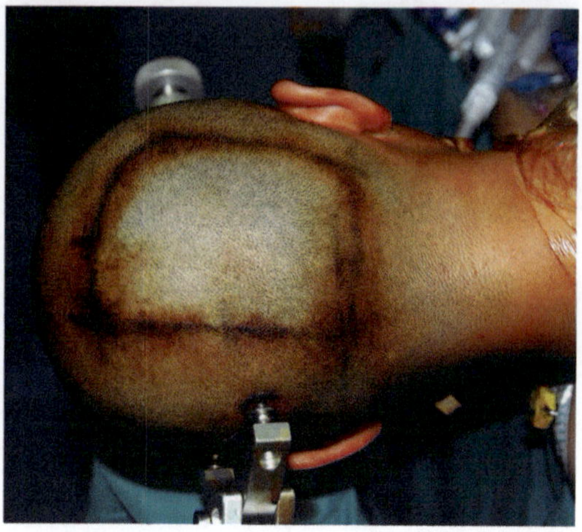

Fig. 27 Lateral decubitus position was used marking parietooccipital "horseshoe" incision

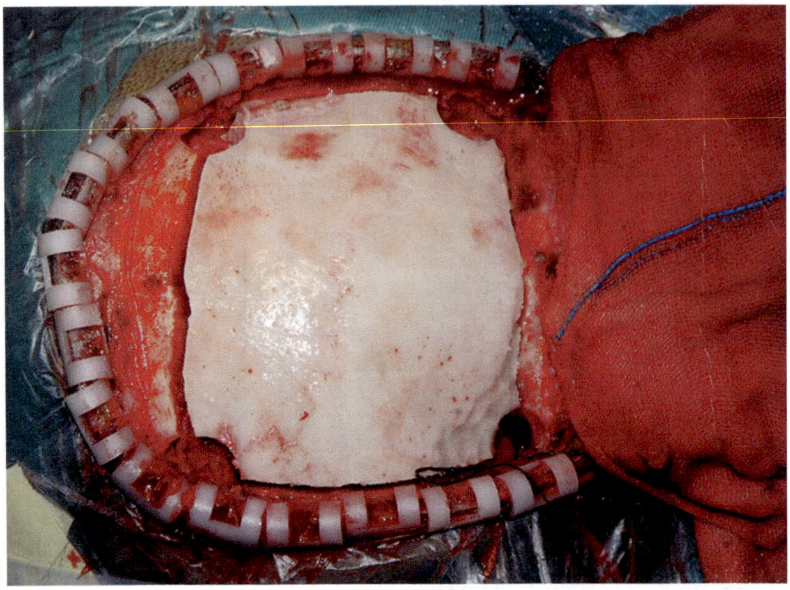

Fig. 28 Four burr holes were made following inferior reflection of muscular flap

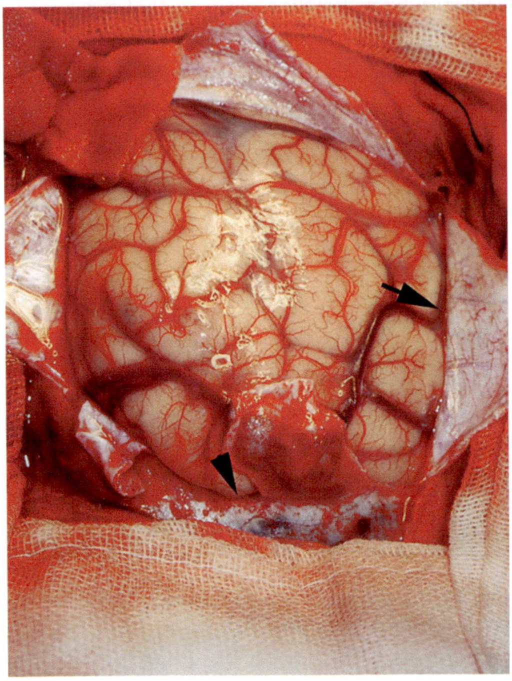

Fig. 29 Dura was opened in a cruciate fashion. Basal aspect of occipital lobe (arrow) and posterior sagittal sinus (arrowhead) was marked

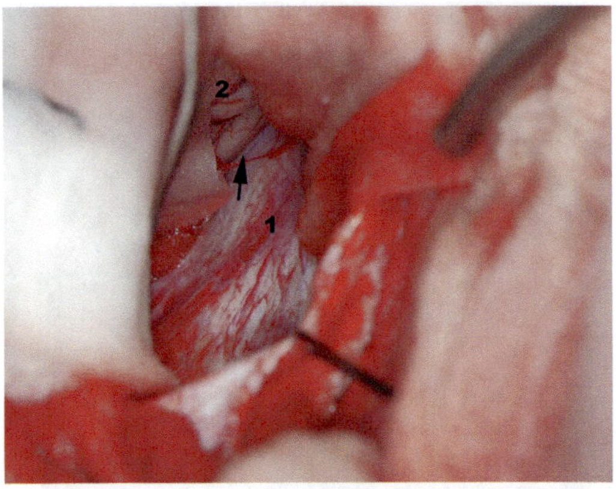

Fig. 30 Occipital lobe was laterally retracted and then quadrigeminal cistern was approached along tentorium edge (arrow). (*1*) Tentorium and (*2*) Lingual lobe

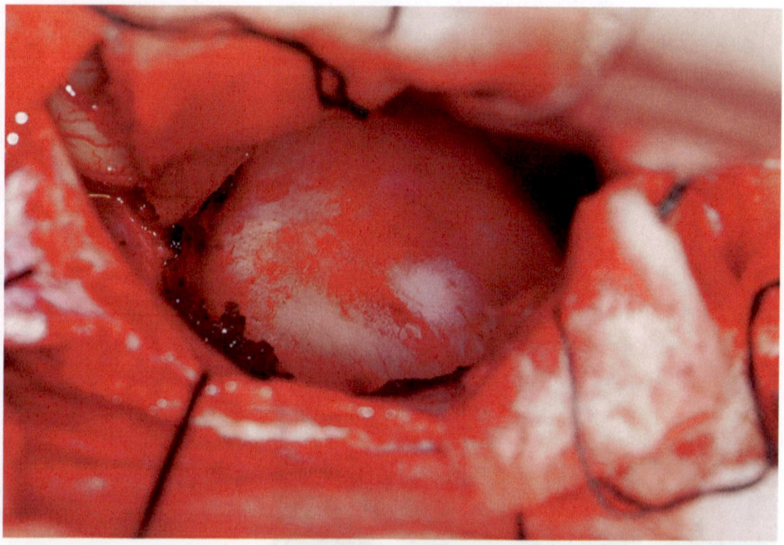

Fig. 31 The vermis tumor was visualized following tentorial opening

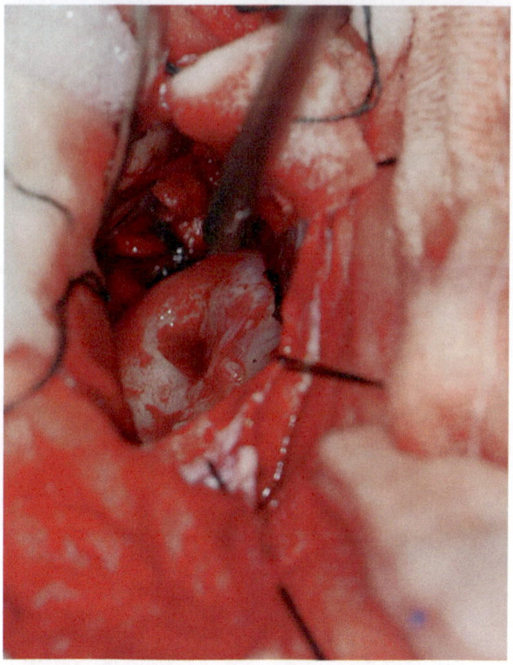

Fig. 32 The mass was resected in a piecemeal manner

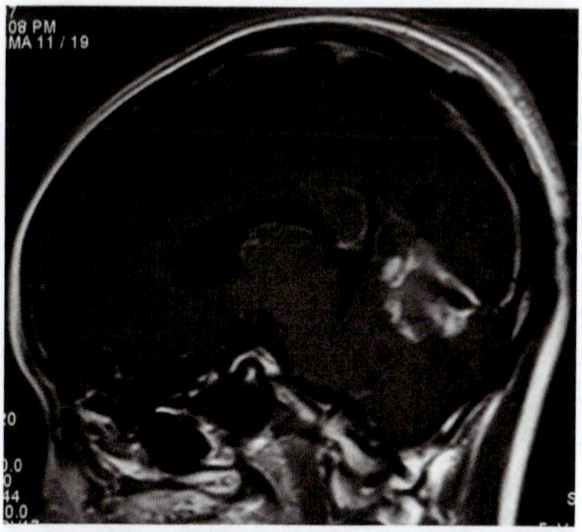

Fig. 33 Post-op sagittal Gd-enhanced T1-weighted MRI confirmed the removal of the tumor

Case 8.3 Germ Cell Tumor of Pineal Region

Clinical presentation:

A 19-year-old male presented with a 2-weeks history of severe headache and vomiting, with gait instability from last week.

Neurological exam:

Papilledema, finger-to-nose test (+), Romberg's test (+)

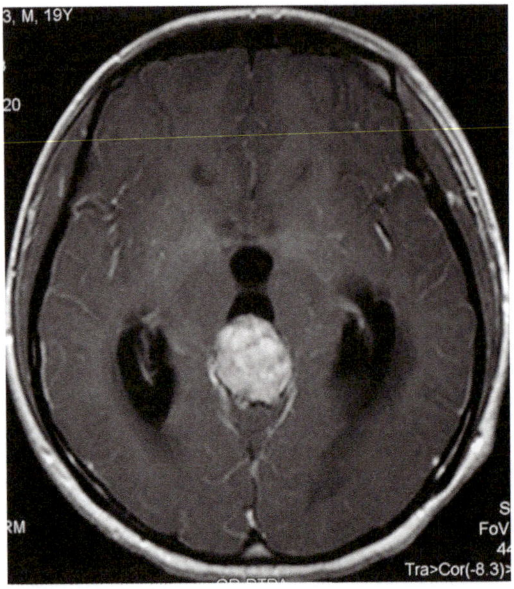

Fig. 34 Pre-op axial Gd-enhanced T1-weighted MRI

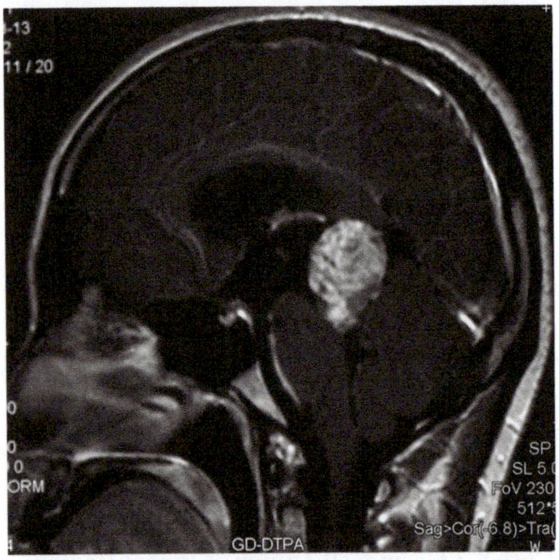

Fig. 35 Pre-op sagittal Gd-enhanced T1-weighted MRI. Pre-op Gd-enhanced T1-weighted MRI showed a solid, well-marginated lesion with heterogeneous enhancement. The cerebral adequate was deformed and obstructed that created dilation of ventricular system

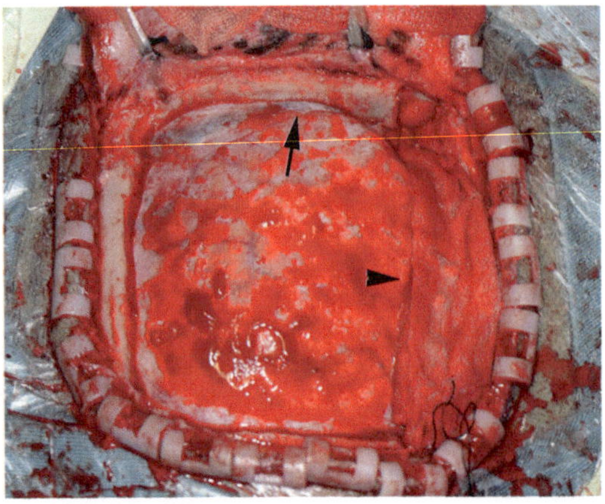

Fig. 36 The right parietooccipital craniotomy was performed and straight (arrow), superior sagittal sinus (arrowhead) were identified

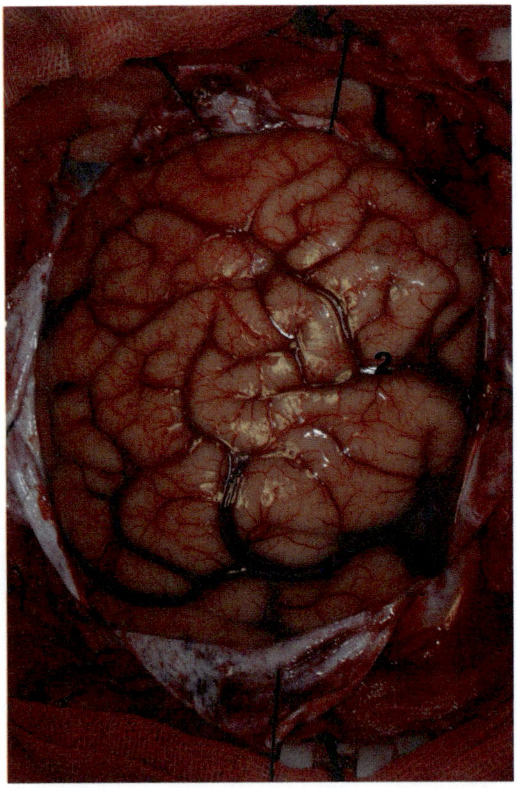

Fig. 37 Dura was incised in a cruciate manner to expose occipital lobe. (*1*) Posterior parietal vein and (*2*) occipital vein

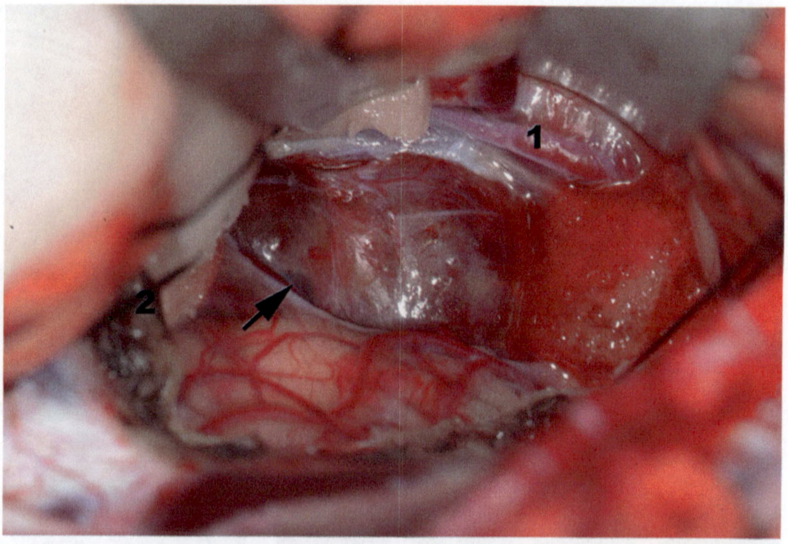

Fig. 38 The basal surface of occipital lobe was retracted superiorly followed by cauterizing the tentorium by outside-in fashion to approach lateral portion of falcotentorial junction. The solid tumor was indicated by an arrow. (*1*) Medial hemispheric branch of superior cerebellar artery and (*2*) cutting edge of tentorium

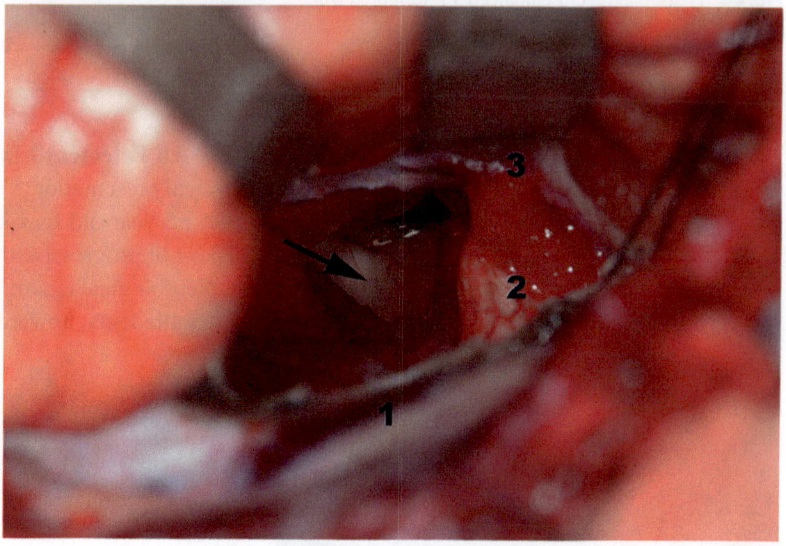

Fig. 39 The posterior wall of 3rd ventricle (arrow) was opened. (*1*) Transverse sinus, (*2*) tectum, and (*3*) superior cerebellar artery

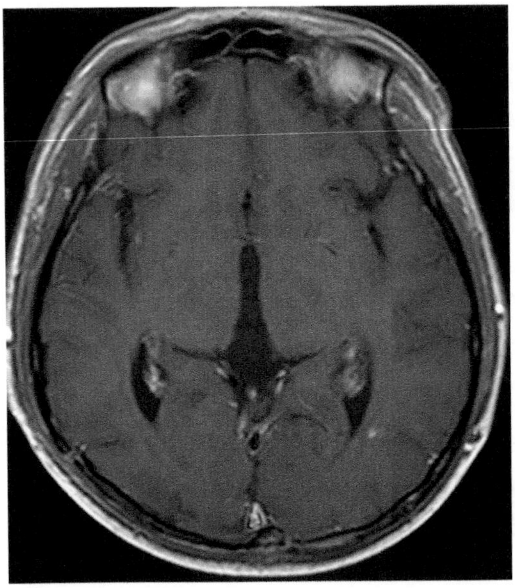

Fig. 40 Post-op axial Gd-enhanced T1-weighted MRI

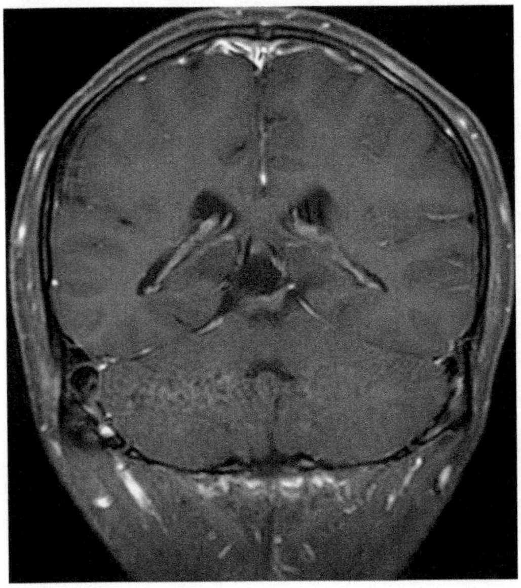

Fig. 41 Post-op coronal Gd-enhanced T1-weighted MRI

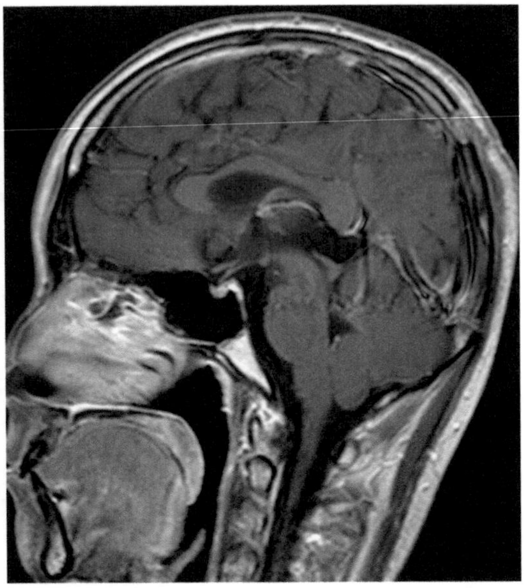

Fig. 42 Post-op sagittal Gd-enhanced T1-weighted MRI. Post-op Gd-enhanced T1-weighted MRI showed total gone of the lesion with intact preservation of deep venous structures.

Case 8.4 Falcotentorial Meningioma

Clinical presentation:

A 55-year-old female presented with a 7-year history of progressive dizziness and headache

Neurological exam:

Papilledema, upward gaze palsy

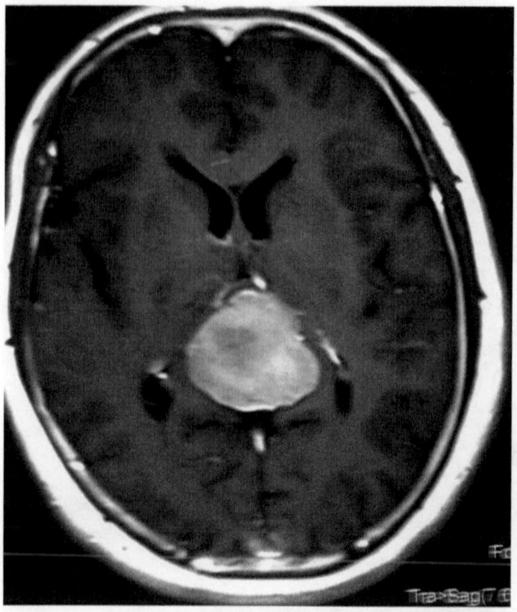

Fig. 43 Pre-op axial Gd-enhanced T1-weighted MRI

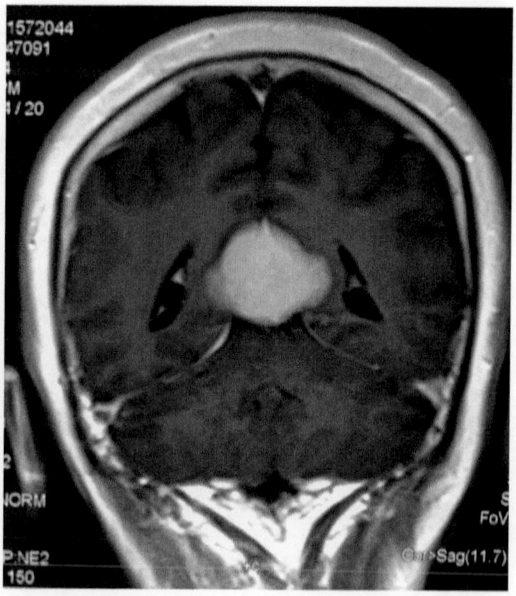

Fig. 44 Pre-op coronal Gd-enhanced T1-weighted MRI

Fig. 45 Pre-op sagittal Gd-enhanced T1-weighted MRI. Pre-op Gd-enhanced T1-weighted MRI revealed a globular solid mass with homogeneous enhancement. This well-marginated lesion was tightly attached with tentorium but has made 3rd ventricle distorted

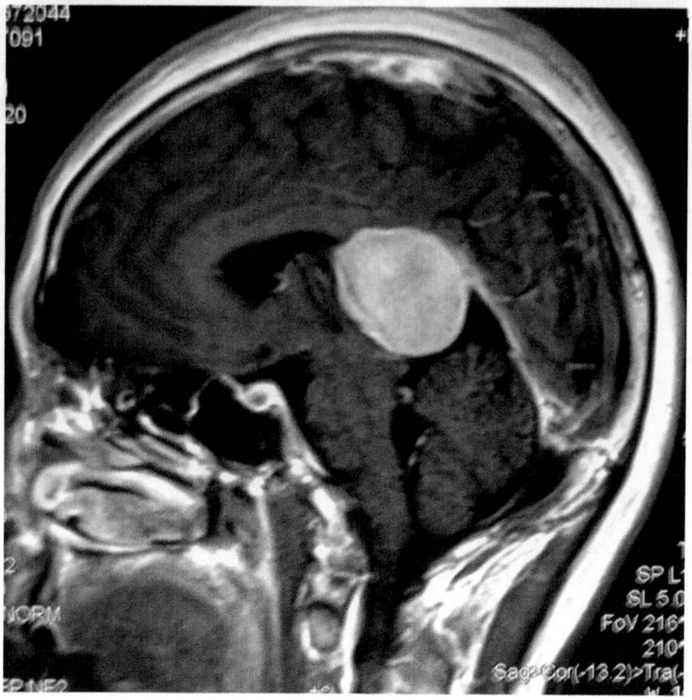

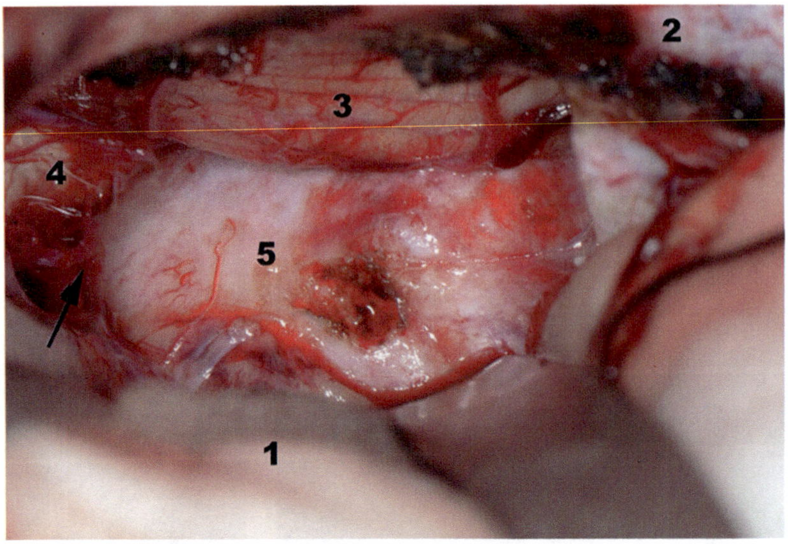

Fig. 46 The basal surface of occipital lobe was retracted superiorly followed by unveiling the nidus and superior cerebellar artery (arrow). (*1*) Basal occipital lobe, (*2*) tentorium, (*3*) quadrangular lobe, (*4*) tectum, and (*5*) tumor

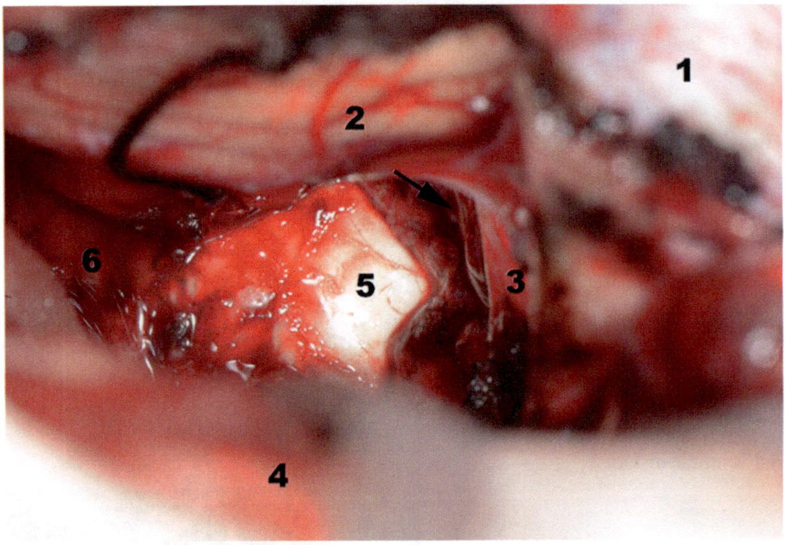

Fig. 47 The vein of Galen (arrow) was seen after tumor removal. (*1*) Tentorium, (*2*) quadrangular lobe, (*3*) straight sinus, (*4*) basal surface of occipital lobe, (*5*) tectum, and (*6*) superior cerebellar peduncle

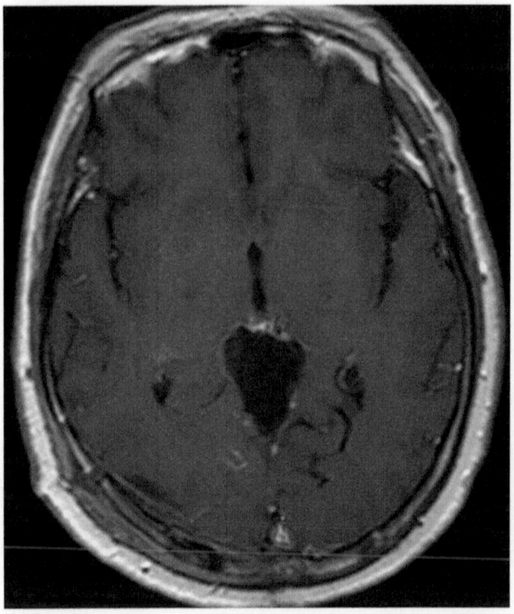

Fig. 48 Post-op axial Gd-enhanced T1-weighted MRI

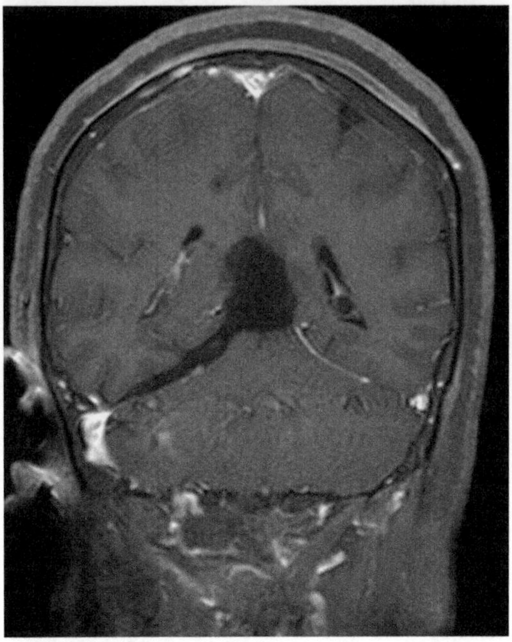

Fig. 49 Post-op coronal Gd-enhanced T1-weighted MRI

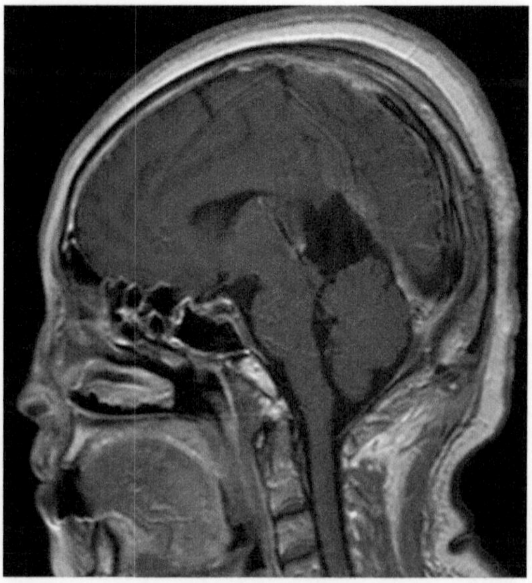

Fig. 50 Post-op sagittal Gd-enhanced T1-weighted MRI. Post-op Gd-enhanced T1-weighted MRI confirmed aggressive resection of posterior 3rd ventricle tumor and good flow through cerebral aqueduct

Case 8.5 Pineal Parenchymal Tumor of Intermediate Differentiation

Clinical presentation:

A 29-year-old gentleman presented with an 11-month history of dizziness and vomiting.

Neurological exam:

Double vision, upward gaze palsy

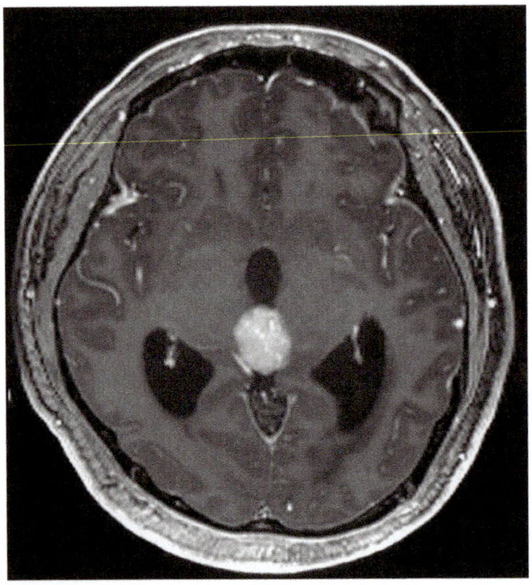

Fig. 51 Pre-op axial Gd-enhanced T1-weighted

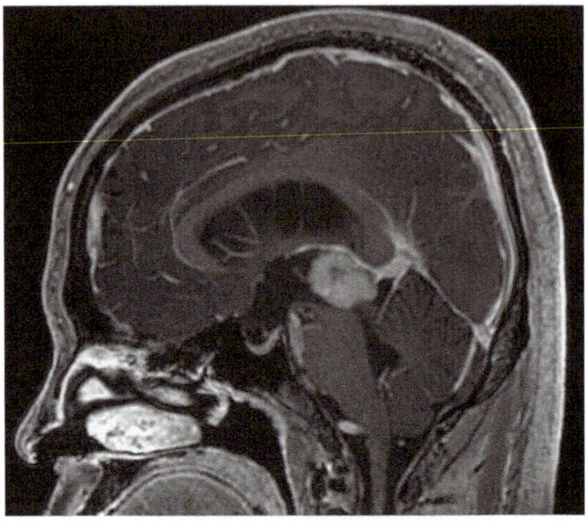

Fig. 53 Pore-op sagittal Gd-enhanced T1-weighted. Pre-op Gd-enhanced T1-weighted MRI demonstrated a globular mass originating from pineal gland and extended into 3rd ventricle resulting in hydrocephalus. This lesion was mostly solid and homogeneously enhanced

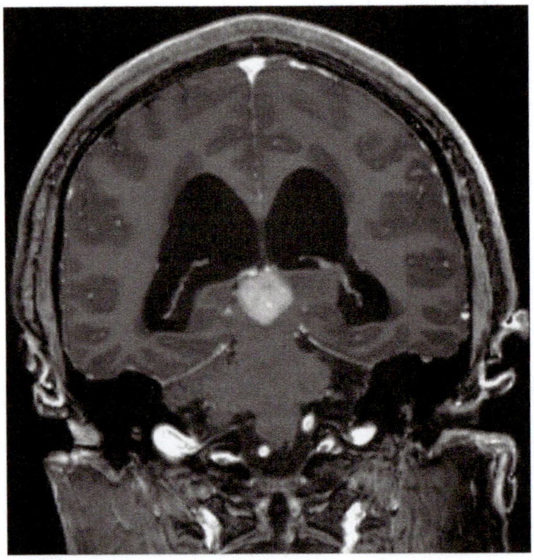

Fig. 52 Pre-op coronal Gd-enhanced T1 weighted

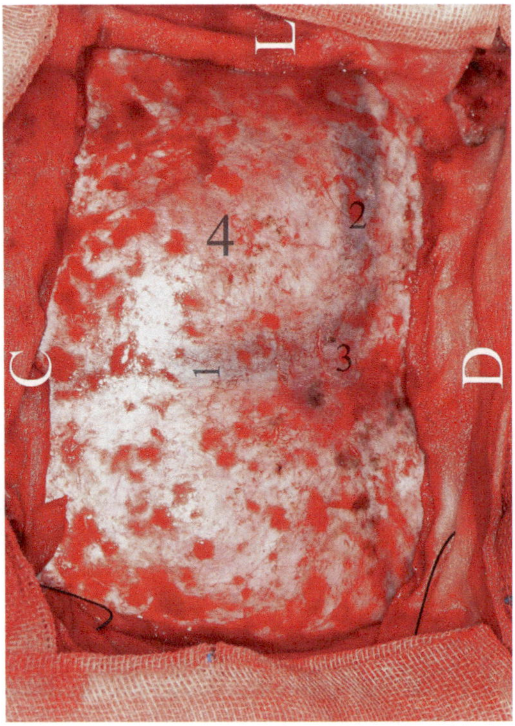

Fig. 54 He was placed in lateral position. Bone flap was created and occipital dura was seen. (*1*) Posterior One-third of sagittal sinus, (*2*) straight sinus, (*3*) torcular, and (*4*) right occipital region

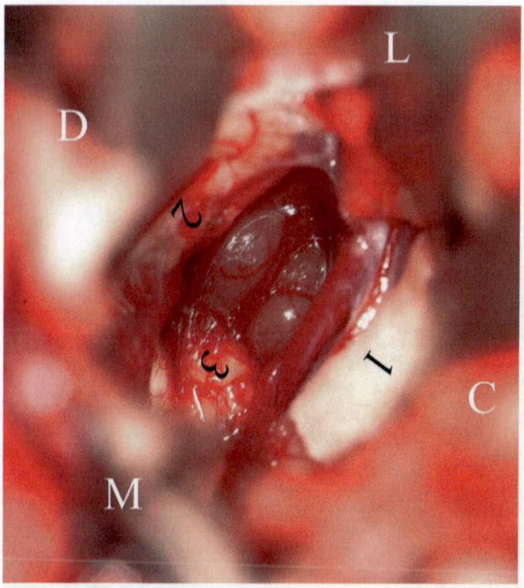

Fig. 55 After incising the tentorium, quadrigeminal cistern was approached and then tumor was visualized with cerebrospinal fluid release. (*1*) Basal surface of occipital lobe, (*2*) quadrangular lobe, and (*3*) tectum

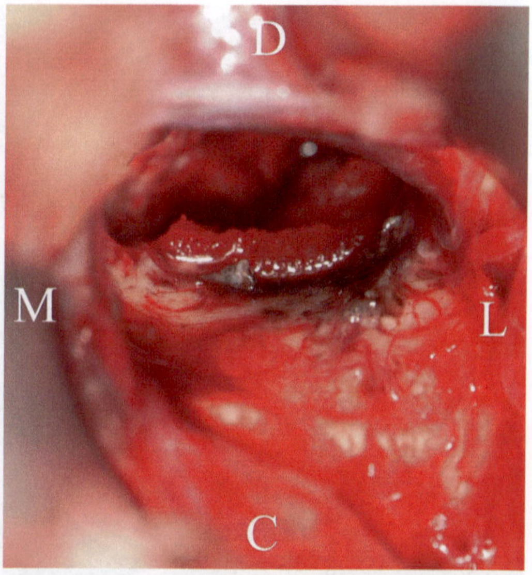

Fig. 56 The 3rd ventricle and cerebral adequate were seen with tumor resection

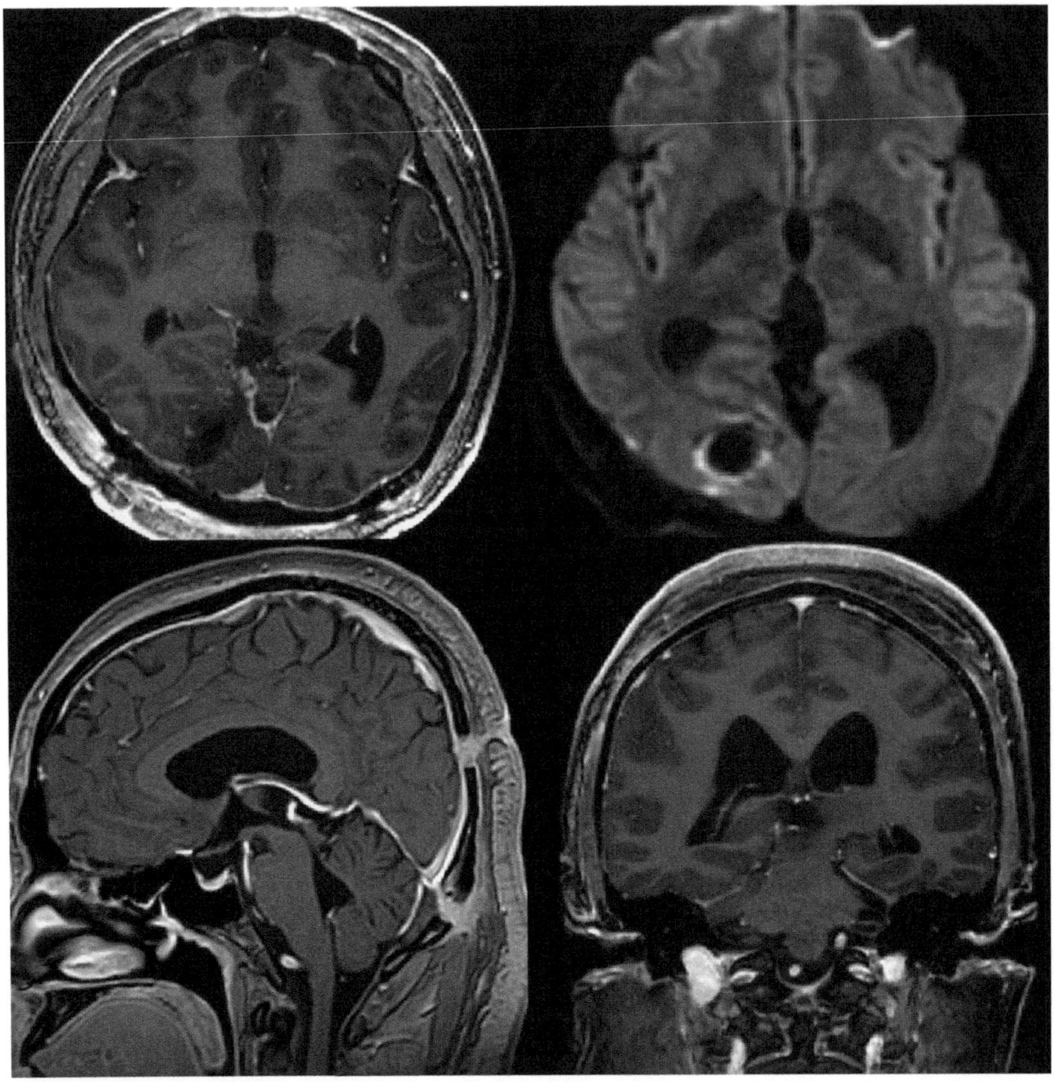

Fig. 57 Post-op MRI demonstrated total resection of the mass without occipital infarction

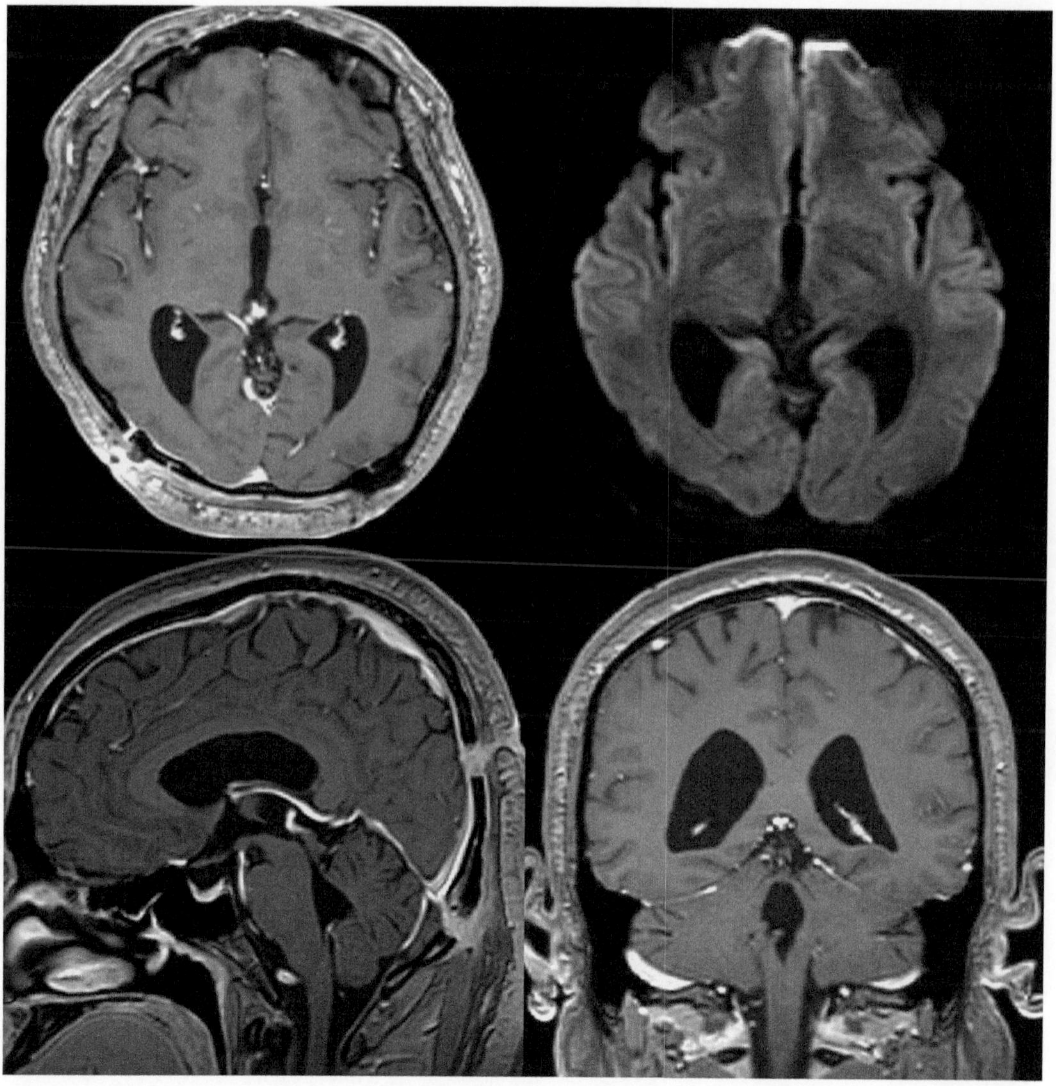

Fig. 58 Post-op MRI at 3-month follow-up showed no occurrence of the tumor

GPSR Compliance

The European Union's (EU) General Product Safety Regulation (GPSR) is a set of rules that requires consumer products to be safe and our obligations to ensure this.

If you have any concerns about our products, you can contact us on ProductSafety@springernature.com

In case Publisher is established outside the EU, the EU authorized representative is:

Springer Nature Customer Service Center GmbH
Europaplatz 3
69115 Heidelberg, Germany

Batch number: 10731407

Printed by Printforce, the Netherlands